ADVANCED AERIAL DEVICES

reported

DURING THE KOREAN WAR

Richard F. Haines

LDA Press, Los Altos, California
P. O. Box 880
94023-0880

ISBN 0-9618082-1-7

Printed in U.S.A.

FOREWORD

After a period of low activity, the mystery of unidentified flying objects is back in the news. Along with a large number of curious readers it is now attracting a new generation of serious students. Many of them are young people who have taken an interest in the most recent American books, films and stories about the phenomenon, without much knowledge or appreciation for its earlier phases. As they start digging into the past and as they gather documentation to feed their curiosity they will find that UFOs have been around for a very long time and that they have been seen not only in America but all over the world.

Good knowledge of these earlier patterns is essential to place current events in the proper perspective.

The period covered in this book, namely the first half of the Decade of the Fifties, should be of special interest to such readers. Dr. Haines has taken the intriguing and unusual vantage point of the Korean War, a conflict that placed thousands of Americans in a faraway land. Will we find that their experiences with flying disks of unknown origin matched those of the folks back home? With the enormous detection and tracking power at their disposal, what did the U. S. Armed Forces learn about the elusive objects? The answers are clearly important for our understanding of the overall phenomenon.

Dr. Richard Haines was the right person to document this period and to ask such questions. An expert in the psychology of perception and a skilled investigator of UFO events, he is as comfortable testing the reactions of pilots to visual stimuli in the laboratories of NASA as he is interviewing witnesses of unusual aerial sightings on a windswept mountaintop. He is one of the few true scientists in this difficult field. He brings to this study an impeccable methodology and he is always careful to separate observation and measurement from illusion and speculation.

A French physicist named Laplace once observed (in his 1812 book *Analytic Theory of Probability*):

> We are so far from knowing all the agents of
> nature that it would be unphilosophical to deny
> the existence of phenomena simply because they
> are unexplained in the current state of our knowledge.
> However, we must examine them with a degree of
> scrutiny which is all the more intense that it seems
> more difficult to admit them.

It is to such "intense scrutiny" that Dr. Haines has subjected the UFO reports made during the Korean War.

Many of the cases he cites are fascinating, but the reader will want to study with special care the sighting near Chorwon in the Spring of 1951 mentioned in Chapter Two, an event in which an entire artillery unit fired at a hovering disk displaying remarkable properties. In my opinion it is one of the most significant reports in the entire literature because of the rich combination of physical and physiological facts it provides.

Many other periods in the tumultuous history of UFO reports should be analyzed in the manner used here by Dr. Haines. Now that he has shown us how to conduct such an analysis, it is my hope that others will undertake this interesting and rewarding task.

Jacques Vallee
San Francisco

ACKNOWLEDGEMENTS

It is indeed a pleasure to thank the many people who helped me in various ways during preparation of this book. First of all, to each of the many eye witnesses of these events goes my gratitude and respect for their courage to report their experiences in the first place, particularly to Mr. Francis P. Wall. It will partially be through such brave acts that we will, one day, discover the core identity of the so-called UFO phenomenon. In addition, the following persons deserve my sincere thanks.

To the Archives Division of the Smithsonian Institution's National Air and Space Museum for permission to reproduce selected photographs of combat aircraft. To John P. Timmerman Vice President of Public Relations for the *J. Allen Hynek Center for UFO Studies* for obtaining part of the interview material in Chapter Two. To Loren Gross for his able assistance in locating some USAF Project Blue Book reports as well as old newspaper clippings. To David L. Black, Director of Public Affairs, Defense Mapping Agency for helpful advice on how to obtain charts from the Korean War period. To Lt. Col. Clayton R. Newell, Chief Historical Systems Division, Department of the Army. To the Chief of Military History at the Center of Military History who provided valuable suggestions on how to obtain further wartime data. To John R. Gerfen, Chief, Army Reference Branch, National Personnel Records Center for copies of wartime unit personnel rosters. Finally, I am grateful to my wife Carol for her expert editorial assistance which made this text far more readable than it otherwise would have been.

LIST OF ILLUSTRATIONS

LIST OF TABLES

CONTENTS

Overview

This book begins with a brief overview of the major historical events of the Korean War period to help set the stage for the UFO sighting reports which follow. Also presented are some important UFO events which took place just before and during the war years. Forty two UFO reports are presented here. More than 63 military pilots, soldiers, ground radar operators, naval personnel, and others covering the period September 1950 to the winter of 1954 contributed to these interesting reports. Six of these cases (14%) involve some kind of electromagnetic effect while another seven (17%) include flight maneuvers by aerial phenomena that rival or excede those of the airplanes that flew in the war. Another six cases (14%) strongly suggest intelligently guided flight control of the UFO relative to the airplane's movements. UFO shape names (number of each in parentheses) include: disc (9), sphere, round, circular (7), oval (2), cigar, Japanese lantern, coolie hat, cylinder, coin, cartwheel, cloud-like. Of the eighteen cases (43%) in which sighting duration is cited, the mean is 6.14 minutes. Fifteen sightings (36%) took place during daylight hours. One report was made during 1950, three in 1951, 24 (57%) in 1952, five in 1953, and one in 1954. Of those that occurred in 1952, seven took place in May and seven in June. Descriptions of the UFOs involved during the May-June 1952 period show a remarkable consistency including such shape names as: oval that is "larger than a MiG", "50 foot diameter", "circular dark object that is flattened on top and bottom", "round", "coin with a 7:1 ratio", "disc with 7:1 ratio and 15-20 foot diameter", "revolving disc". Although the U.S. Air Force's official conclusions for many of these cases was a lighted enemy balloon, most of the sighting data do not support this explanation. It is quite clear that the reported phenomena are closely similar in all major respects to other UFO cases both before and after the Korean War. There is virtually no evidence of hostile intent shown by the UFO(s) during any of these close aerial encounters. Is it possible that others were only interested in how wars are fought on Earth?

Chapter 1

The Stage is Set

Barely 59 months after the Japanese signed an unconditional surrender on September 2, 1945, a relatively minor civil war broke out on the Korean Peninsula between the Democratic Republic of (South) Korea and the communistic Democratic People's Republic of (North) Korea. Some historians point out that what started as a civil war limited to a then third rate economic and strategic nation; escalated rapidly into an undeclared training ground for new weapons systems, advanced strategic planning, and armed forces drawn from more than seventeen other nations fighting on the side of freedom and democracy. The enemy comprised two nations (North Korea, Red China) with the Soviet Union standing in the background, supplying training personnel and materiel. Many claimed that the Korean conflict was only an inevitable consequence of the larger "cold war" that was raging between America and the Soviet Union.

America's real military strategy in Korea was "...to ensure that it did not grow into World War 3. This meant that political leaders (in Washington) were in charge of the war strategy rather than the military leaders (in the field)." (Anon., pg. 3-38, 1986).

The Korean peninsula (see Figure 1) was supposed to become an independent nation during WW2 (according to the super powers United States, Soviet Union, Great Britain, and France). "When the war with Japan was over, there were no US troops in Korea. Since it would take some time to move troops into the peninsula, the United States asked that the Soviets accept the surrender of all Japanese troops north of the 38th parallel in Korea and the United States all those south of the parallel. The division was supposed to be a temporary measure, but the Soviets began to treat it as a permanent boundary, and they took control of North Korea. The Soviets did not want to see Korea become a free nation and come under Western influence." (Anon, 1986)

Figure 1

The Korean Peninsula

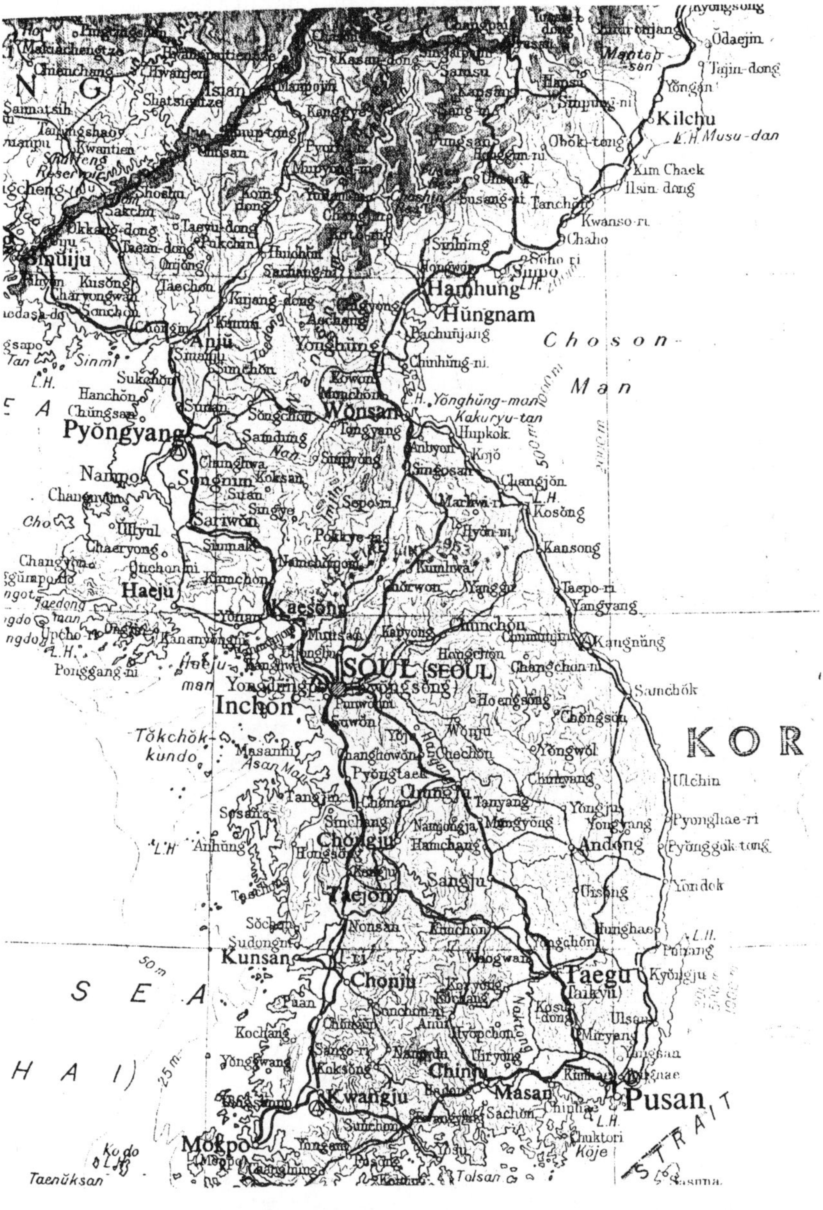

But our concern is not so much with politics as it is with the state of advanced warfare technology at this time. What did the United States Air Force have in its arsenal during this war? What were the Russians flying? Were there other nations with advanced flying craft that did not have wings but could fly higher and faster than anything either the Russians or the Americans had? The answer is a simple NO!

Of special note is the fact that recently uncovered, seemingly authentic U.S. government documents have indicated that our government had in its possession at least one highly advanced spacecraft not from this planet which was allegedly recovered in 1947 in New Mexico (note 1). If this is true and scientists had succeeded in understanding how to duplicate its advanced technology, then some or all of the UFO reports coming from Korea could represent an American technological development. Then one would have to ask: (1) why these disks were not used in a more aggressive way to help win the war, (2) why these disks never crashed or were never recovered by anyone, (3) why our Air Force has continued to spend millions of dollars (since the Korean war) on turbo-jet engines and swept-wing airplanes rather than on self-propelled metallic-surfaced oblate spheroid (discs), and (4) why we have seen no UFO like aerial objects fighting in the Viet Nam war to help the U.S. forces win. If the aerial objects to be described are not American, Soviet, or from another nation on earth then an intriguing possibility exists; viz., they represent an alien technology.

One of the interesting things about the Korean "Conflict" as it was called at the time, was the role played by the world's two super powers, each supplying their respective surrogate Korean armies in the field, at least until November of 1950 when soldiers and materiel from Communist China flooded across the Yalu River into North Korea. What had started out as a localized peace keeping action by United Nations troops suddenly presented the spectre of all-out war with China, the most populous nation on the earth. A nuclear sword swung tenuously above both sides of the conflict.

The Soviets had largely equipped North Korea's army and its air force. It was, perhaps, partly a field training exercise for them to see how well their planes and tanks, their ammunition and other war materiel would fare in the extremely cold weather and in the hands of less well trained soldiers. On June 25, 1950 the North Koreans had flooded across the 38th parallel, the former boundary with their cousins to the south, and quickly overran the smaller and less well prepared South Korean units. With the Soviet delegate voluntarily absent, the United Nations Security Council in New York City invoked military sanctions against North Korea (June 27, 1950) and formally requested its member states to give whatever aid they could to South Korea. The die had been cast.

Poured into this already fiery hot mold were soldiers from the following countries:

United States of America	Italy (not a U.N. member)
Australia	Netherlands
Belgium	New Zealand
Luxembourg	Philippines
Canada	South Africa
Columbia	Thailand
Ethiopia	Turkey
Great Britain	
Greece	

Many hundreds of thousands were to be be maimed or killed in the fray. The mold was severe and unforgiving as it is in any war to those who must do the fighting. A total of 25,604 U.S. servicemen were killed in this "conflict" with another 137,051 listed as casualties. South Korea lost 415,004 soldiers with more than 1,312,800 casualties. The other U.N. participating nations lost 3,094 men with over 16,500 casualties. It has been estimated that the communist's casualties were about two million (Morse, vol. 15, 1969).

The free world's armed forces were unified under United Nations command. It was headed by General Douglas MacArthur, Commander in Chief, Far East Command (FECOM). General MacArthur reported to the Joint Chiefs of Staff concerning all of the U.S. forces (Momyer, 1978). Figure 2 illustrates the command structure for U.S. air operations in Korea in 1950. What it doesn't show is General MacArthur's failure to establish an army component command since "He reserved to himself the roles of the Far East Command structure." (Ibid., pg. 53) Both the U.S. Navy and Air Force set up respective components named the Naval Forces Far East and Far East Air Force commands, respectively. Both had staffs manned so as to direct their respective forces throughout the area of MacArthur's responsibility. Whether or not this short-coming (not having a balanced staff represented by all of the armed services) contributed to how UFO sighting reports were handled remains to be seen.

The U.S. Navy's Task Force 77 operated off the Eastern coast of Korea in the Sea of Japan with its aircraft carriers providing interdiction of enemy aircraft, bombing support, and close air support for Marine operations up to 70 miles inland along the entire length of North Korea's coast.

Reference to Figure 2 shows that the Commanding General of the 5th Air Force was located in Korea and exercised operational control of Marine aircraft as well as coordinating bomber strikes with all other forces such as flak suppression and fighter support.

Figure 2

Command Structure for U.S.
Air Operations in Korea

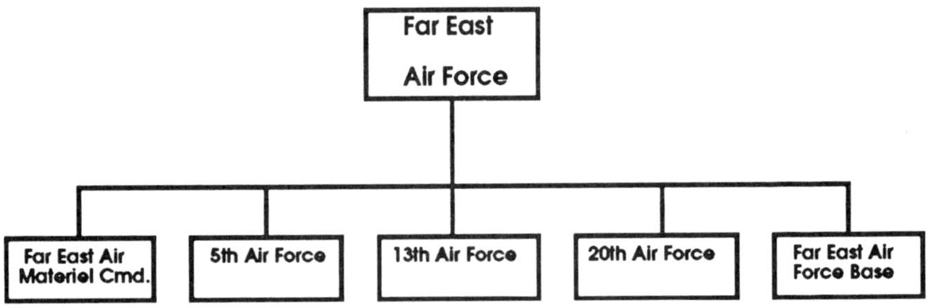

U.S. Air Power During the Korean War:

But what about the Air Force participants? Who were they? What units were called up for service so soon after the Second World War had ended? The Far East Air Force (FEAF) consisted of these units

5th Air Force
13th Air Force
20th Air Force
Far East Air Materiel Command
Far East Air Force Base

The Far East Air Force Command (FEAF) was under the command of Lieutenant General George E. Stratemeyer (1890-1969). On October 8th 1950 he requested that he be given full operational control of all air units. This meant that he would be able to fully coordinate the Air Force's mission with those of other ground forces; even specifying the amount of forces to be deployed, the type of munitions, the time on and off targets, and the controlling agencies.

FEAF chose all their targets (both for the Air Force and Naval carrier-based aircraft) by means of a "targeting committee" that was composed of Navy and Air Force representatives. This coordinated approach had proven itself in North African air operations where there were little or no industrial targets and other targets required less force to destroy or neutralize.

As indicated above, the 5th Air Force was coordinated by the Far East Bomber Command and, in turn, coordinated fighter escort. Air route planning to and from targets was the joint responsibility of the 5th Air Force and Far East Bomb Command. B-29 bombers were used extensively during the war (Figure 3). The Far East Bomber Command consisted of three B-29 groups drawn from the Strategic Air Command (SAC).

Following are different airplane models flown by U.S. pilots during the Korean War: (Maximum speed [mph] is given in brackets)

F-80 (Shooting Star)[543]
F-84 (Thunder jet)[622]
F4U-4B (Marine)[450] (see Figure 9)
F7F-3N [427]
F9F-2 (Panther jet) (note 2)[625]
F-94 [600] (see Figure 7)
AD Skyraider (Navy, propeller-driven aircraft)(note 3)[320]
P-51 (Mustang)[370]
F-86 (Sabrejet)[680] (see Figure 4)
T-6 [205] (see Figure 10)
B-26 (2 engine bomber)[282]
B-29 (4 engine heavy bomber)[358] (see Figure 3)
C-54 (troop transport)[274] (see Figure 8)

Figure 3

B-29 Bomber in Flight
(Reproduced by permission of the
National Air and Space Museum
Smithsonian Institution)

The F-100 Super Sabre flew for the first time in 1953 but was not used in the Korean War.

North American Aviation's all-weather F-86 Sabrejet was the nearest combat fighter America had to the MiG-15 in most operational respects. It was the "...best aircraft the U.S. had during the Korean War" (Anon., pg. 3-41, 1986). It was 41 feet long with a wing span of 37' 1". Its loaded weight was 16,500 pounds. The F-86 had a maximum speed of over 660 mph and a service ceiling of about 50,000 feet. Its armament consisted (model E) of six 50 cal. machine guns in the nose. There also were provisions for 16-127mm rockets under the wings and two each 1,000 pound bombs or two each 2,000 pound bombs in lieu of auxiliary fuel tanks. Three prototypes were ordered in May 1945; the XP-86 flew for the first time on October 1, 1947.

Figure 4

F-86 Sabre Jet
(Reproduced by permission of the National Air and Space Museum Smithsonian Institution)

In November 1950 the 4th Fighter Interceptor Group arrived in Japan with its F-86As and soon operated out of Kimpo air base in South Korea (Jackson, 1979). Regarding air operations, Braybrook (1987) said that F-86s generally "...flew in sections of four aircraft, up to eight sections together, and between 35,000 and 45,000 feet...depending on model. The MiG-15s crossed the border at around 50,000 feet in 'trains' of 60 to 80 aircraft." He goes on to point out that most F-86 kills of Communist aircraft were made without radar; most used a fixed gunsight and approached the enemy from the enemies' 6 o'clock position.

By the close of the war there were seven U.S. fighter wings in Korea with

297 F-86s and 218 F-84s. Sabre operations peaked at 7,696 sorties in June 1953 and an average of 26 sorties per aircraft for that month (Braybrook, 1987). Sabres destroyed 810 enemy aircraft, 792 of which were MiG-15s. Seventy eight U.S. aircraft were lost.

U.N. airplanes provided air cover for U.N. ground forces. Two weapons in particular proved to be the best new weapons in Korea, napalm and aerial rockets. The rockets had the destructive force of a 105mm cannon shell. Napalm bombs were 110 gallon tanks of jelled gasoline which exploded in fire over an area 250 feet long and 80 feet wide.

What could an American pilot do if he chased a Soviet MiG airplane north over the Red Chinese border? The FEAF commander was given guidance from Washington. So called "hot pursuit" was authorized under some conditions, "...but attacks against aircraft taking off from bases across the Yalu (river) were not." (Ibid., pg. 56) It is interesting to note that none of the UFO sighting reports presented here includes a "hot pursuit" very far because the unidentified aerial craft *almost always* outperformed the pursuing jet airplanes.

Soviet Air Power During the Korean War:

This is a list of some of the airplanes flown by the air forces of the North Koreans, the Chinese Communists, and the Soviet Union.

La-9 (single engine, propeller driven fighter) [430]
La-11 (single engine, propeller driven fighter) [420]
Ilyushin 10 (single engine, propeller driven fighter) [300]
YAK-9 (single engine, propeller driven fighter)
YAK-15 (single engine, propeller driven fighter)
MiG-15 (single engine, jet fighter) [680] (see Figure 5)
TU-2 (bomber) [340]

It is instructive to note that at this time the Soviets had over 15,000 MiG model 15s available (Nowarra and Duval, pg. 168, 1972). Stockwell (1956) estimates the number to be from 12,000 to 15,000. The Red Chinese Air Force supposedly had about 1,000 MiG-15s as they entered the Korean War. This jet fighter interceptor was affectionately code named "Fagot" by N.A.T.O. officials. Braybrook (1987) points out that the number of Communist aircraft peaked at about 1,800 (950 MiGs) with over 400 parked at one airfield in North Korea.

The MiG-15 measured only 36' 4" long, with a wing span of 33' 1" (Figure 5). It weighed 8,316 pounds empty and could carry a payload of 5,907 pounds in addition to its single pilot. Its gross weight was 11,264 pounds. Its top speed was about 680 mph (Mach limit = 0.89) at sea level and had a service ceiling of 51,000 feet and a range of 1200 miles with underwing fuel tanks. It stalled at (or possessed a minimum air speed of) 109 mph. The MiG-15 carried one 400 rounds per minute, 37mm cannon and two 23mm caliber cannons in addition to two each 990 pound bombs. This fighter entered the war on November 1, 1950 when a flight of six aircraft attacked Air Force Mustangs south of the Yalu River without doing any damage. Soviet units regularly flew combat duty over Korea in conjunction with Chinese Communists and North Korean formations (Jackson, Pg. 94, 1979).

The U.S. Air Force Junior ROTC publication "Aerospace Science: History of Air Power" (Anon., 1986) states that the MiG-15 "...was faster, more maneuverable, could climb faster and higher, and possessed more firepower than the F-80, F-84, or the Navy F-9F (sic) fighters. In fact, the MiG-15 even had the edge, at high altitude, over the F-86 Sabrejets which were the best aircraft the United States had during the Korean War." Because of superior pilot skill by U.S. pilots, nine MiGs were shot down for every U.S. aircraft.

The La-9 was a Soviet designed and built fighter, code named "Fritz". It possessed a maximum speed of 430 mph at sea level and a service ceiling of 35,600 feet. Its wing span was 34' 9" and was 30' 2" long. This piston-driven propellor airplane was in service until the 1948 - 1950 period.

Another Soviet fighter that was used in combat was the La-11, code named "Fang". It is comparable in design to Republic's P-47N "Thunder-bolt." Delivered in early 1946, this single seat fighter interceptor was only 28' 6.5" long with a wing span of 31' 10". Its top speed was about 420 mph with a service ceiling of about 34,000 feet. It flew in Korea with Chinese and North Korean markings (Jackson, pg. 76, 1979). It carried three each 20 or 23 mm cannons.

A large number of UFO sighting reports are presented in the pages to follow. If these UFO were enemy weapons of war: (1) why would they continue to be used during the truce period? (2) why would the U.S. not be able to identify them more definitively? and (3) why weren't they used more effectively by the enemy during the actual conflict? There were no UFO reports found which demonstrated clearly hostile intent toward U.S. personnel on the part of the unusual aerial phenomena.

Figure 5

Captured Soviet Made MiG-15
(Reproduced by permission of the
National Air and Space Museum
Smithsonian Institution)

Military and Political Events:

It is important to have some idea of the major military and political events which took place before and during the Korean Conflict. A number of them are listed in Table 1. It may be significant that the first UFO sighting report was *not* made until September 1950. It took place 100 miles south of the Yalu River. Twenty two separate sightings occurred after the truce negotiations were underway. This is interesting in light of the fact that open hostilities had ended and yet clearly defined UFO reports continued to come in from observers on the ground and in the sky.

Table 1

Historical Events Surrounding the Korean War
(See Key for abbreviations)

------ ***Historical Background*** -----

Jan. 20, 1945	Harry S. Truman begins Presidency
July 16, 1945	World's first Atom bomb detonated near Alamagordo, New Mexico
Aug. 5, 1945	Atom bomb exploded over Japanese city of Hiroshima
Aug. 15, 1945	Soviet Union takes control of NK military government until Dec. 26, 1948
Sept. 8, 1945	USA takes control of SK military government
March 11, 1948	Key West Agreement (James Forrestal, Secy. of Defense) assembled Joint Chiefs of Staff at Key West to decide "who will do what"
Early 1949	Soviets withdraw all troops from NK; USA does same from SK except for small group of military advisors (withdrawn later that year)
July 1949	President Truman signs North Atlantic Treaty
Dec. 22, 1949	Prototype flight of F-86E in Los Angeles
1949-1950	NK tries unsuccessfully to take control of SK through insurgency operations

------- ***The Civil War Begins*** -----

June 25, 1950	NK army invades SK at nine different points
June 27, 1950	U.N. declares official sanctions against NK. President Truman orders General MacArthur to use air and naval forces in defense of SK
June 28, 1950	Seoul (capital of SK) falls to NK invaders
July 1950	All U.N. forces retreat to perimeter-defense line about 50 miles around Pusan
Sept. 15, 1950	Amphibious landing made at Inchon. First contact of U.S. forces of X Corps with NK forces 200 miles north of the NK defensive positions
Sept. 26, 1950	Seoul recaptured
Oct. 20, 1950	U.N. forces move north across 38th parallel and capture the capital of NK (Pyongyang). Some units move

north of the Yalu river, the national boundary between NK and China

Nov. 1950 Red Chinese army units 850,000 strong, cross into NK to fight against U.N. troops

Nov. 24, 1950 MacArthur orders an "end-the-war" offensive. A massive Chinese counteroffensive almost immediately cancels this thrust

Nov. 26, 1950 Red Chinese soldiers cut the escape route of over 200,000 U.N. soldiers and marines who are evacuated by ship from the port of Hungnan

Dec. 5, 1950 The Chinese hoards sweep south to recapture Pyongyang

Jan. 4, 1951 The Chinese recapture Seoul in first major offensive

Feb. 22, 1951 U.N. Command initiates "Operation Killer" along a broad front well south of Seoul and pushes north with superior firepower

Feb. 1951 Red Chinese make advances in second major offensive

March 14, 1951. Seoul is recaptured by U.N. troops

April 8, 1951 President Truman sends orders relieving Gen. MacArthur of his command. MacArthur had publically advocated direct attacks against the communists in Manchuria, an act considered to be insubordination toward the President and U.S. Congress

April 21, 1951 Gen. MacArthur leaves FEAF command. Gen. Matthew Ridgway (1895-1971) given command of FEAF

April-May 1951 Red Chinese mount third major offensive

----- *Attritive Phase of the War Begins* -----

April 22, 1951 U.N. troops occupy positions just north of 38th parallel along a line that remained almost constant for the remainder of the war

The battle field strategy remained to inflict maximum personnel loss along the fixed battle front and from the air. This approach could not drive the enemy from the field and could never result in total victory, like that achieved in WW2

May 1951 General Van Fleet (Eighth Army Commander) orders a huge coordinated counteroffensive

May 9, 1951 Largest air strike of the war. Over 300 fighters and fighter-bombers attack Sinuiju near the Yalu River

July 10, 1951	Truce negotiations begin at Kaesong between U.N. representatives and Communist commands
Oct. 1952	Negotiations break down over one final principle (i.e., prisoners of war should not be returned to their respective armies against their wills)
Nov. 1, 1952	U.S. explodes its first Hydrogen bomb at Eniwetok Pacific Proving Grounds (Operation Ivy)
Nov. 4, 1952	Dwight D. Eisenhower elected 34th President of the USA
Dec. 1953	President elect Eisenhower visits Korea
April 1953	Negotiations resume
July 27, 1953	Truce agreement signed at Panmunjom
Aug. 1953	Soviet Union explodes a thermonuclear weapon

Abbreviations:
SK = South Korea; NK = North Korea;

The historical events cited in Table 1 are political and military in nature. But what about UFO happenings? There were many other events going on at the same time which should be kept in mind as the war raged in Korea. Some of the more prominent events are listed in Table 2.

Table 2

Historical UFO Events

July 1947	U.S. Air Force begins to study UFO reports seriously after receiving numerous reports by pilots and others in America
Sept. 23, 1947	Chief of the Air Technical Intelligence Center (ATIC) prepares letter to Commanding General of the Air Force stating that it is ATIC's opinion that UFOs are real and urges that a permanent project be established to study them
Jan. 22, 1948	Project Sign (also known as Project Saucer) established at Wright-Patterson AFB, Ohio
Sept. 1948	Top Secret "Estimate of the Situation" prepared by ATIC; sent to A.F. Chief of Staff, Gen. Hoyt S. Vandenberg; returned for more proof; later declassified and burned! (Hall, pg. 106, 1964)
Feb. 11, 1949	U.S. Air Force UFO Project renamed "Project Grudge"
Dec. 27, 1949	Project Grudge final report released; all sight-

ings explained away as delusions, hoaxes, and crackpot reports. Termination of the project is announced

Sept. 15, 1951 A Pentagon general requests briefing on Project Grudge findings by Lt. Jerry Cummings and a Lt. Col. from ATIC; orders given to set up a new study project. Early reports of UFO in Korea very likely figured in this request

Sept. 1951 Capt. Edward J. Ruppelt appointed chief of UFO study activity (supported by: Lt. Bob Olsson, Lt. Henry Metscher, Lt. Andy Flues, and Lt. Kerry Rothstien)

March 1952 Project Blue Book (code name) officially established at ATIC

April 1952 Life Magazine publishes major article "*Have We Visitors from Space?*" Hall (1964, pg. 107) suggests it was inspired by several top Air Force officers in the Pentagon

AF Letter 200-5 issued giving Project Blue Book fuller, direct access to pilot (and other) sighting reports

July 1952 Newly established Aerial Phenomena Research Organization (APRO) publishes first issue of the APRO Bulletin

Civilian Saucer Investigation (study group) of Los Angeles, California is founded (Jacobs, pg. 84, 1975)

Aug. 1952 A USAF study of UFO maneuvers begins; emphasis is on possibility of intelligent UFO control

Jan. 14-17, 1953 A.F. (with C.I.A. according to Hall; Ibid.) convenes top scientists to study all available UFO evidence (Robertson Panel). Maj. Dewey Fournet presents evidence and conclusions that UFOs are of interplanetary origin

Jan. 17, 1953 Panel concludes its review without it being made public. (Hall notes that "since then, two conflicting versions have been released")

Dec. 1953 Joint Chiefs issue "Joint-Army-Navy-Air Force Publication (JANAP) 146" entitled "Canadian United States Communications Instructions for Reporting Vital Intelligence Sightings"

It is very likely that the numerous high quality U.S. military sightings of UFOs from the Korean War zone contributed significantly to continuing Project Blue Book by the U.S. Air Force.

Notes:

1. William Moore presented this astonishing information during the 1987 annual meeting of the Mutual UFO Network held on June 26-28 at The American University, Washington, D.C. He distributed an eight page report entitled "Briefing Document: Operation Majestic 12 Prepared for President-Elect Dwight D. Eisenhower: (Eyes Only), 18 November, 1952" which allegedly documents this.

2. First saw action in Korea on July 3, 1950 when an F9F-2 shot down a Mig-15.

3. First saw action in Korea on July 3, 1950.

Chapter 2

GI Fires Upon an Unidentified Aerial Object

Following is a transcript of an interview between Mr. Francis P. Wall, a private first class in the U.S. Army during the Korean War and Mr. John Timmerman who is Project Manager for the photo exhibit of the *J. Allen Hynek Center for UFO Studies* in the Fall of 1987. It has not been edited in any way. This interesting event began at dusk and ended at about 9 pm local time.

"This event that I am about to relate to you is the truth, so help me God. (Note 1) It happened in the early Spring of 1951 in the country of Korea. We were in the Army infantry. I was in the 25th Division, 27th Regiment, 2nd Battalion, "Easy" Company (note 2). We were in what is known on the military maps as the Iron Triangle, near Chorwon. We were to the left of Chorwon, just across the mountain ridge from this city - town - whatever you want to call it. It is night. We are located upon the slopes of a mountain, between the fingers of a mountain as they as they run down toward the valley below where there is a Korean village. Previously we have sent our men into this village to warn the populous that we are going to bombard it with artillery. Upon this night that I'm talkin' about, we were doin' just that. We had aerial artillery bursts comin' in. And we suddenly noticed down, with the mountains to our backs, we noticed on our right-hand side what appeared to be a jack-o-lantern come wafting down across the mountain. And at first no one thought anything about it. So we noticed that this thing continued on down to the village to where, indeed, the artillery air bursts were exploding. And we further noted, by the way, it had an orange glow in the beginning, we further noticed that this object would get right into... it was that quick that it could get into the center of an airburst of artillery and yet remain unharmed. And, subsequently, this time element on this, I can't recall exactly, I would say anywhere from, oh, forty-five minutes to an hour all told." Figure 6 is a sketch of the object made for the author by the witness on May 10, 1989.

Figure 6

Eyewitness Sketch of UFO
as Seen From the Side During its Approach

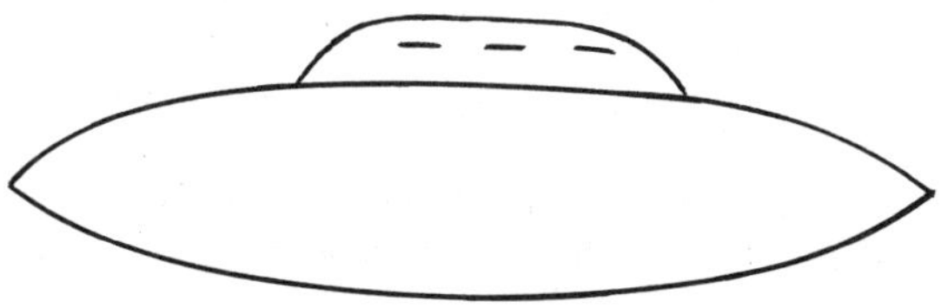

"But then this object approached us. And it turned a blue-green brilliant light. It's hard to distinguish the size of it, there's no way to compare it. It pulsated. The light, that is, was pulsating. It wasn't, ah, regular. Alright, this object approached us. I asked for and received permission from Lt. Evans, our company commander at that time (note 3), to fire upon this object, of which I did with an M-1 rifle with armor-piercing bullets, or rounds in it. And I did hit it. It must have been metallic because you could hear when the projectile slammed into it.

"Now why do you say, why would that, ah, bullet damage this craft if the artillery rounds didn't? I don't know unless they had dropped their protective field around 'em, or whatever. That this, ah, technology envisions, that they had to protect it. But the object went wild and it... the light was goin' on and off and it went off completely once, briefly. And it was moving erratically from side to side as though it might crash to the ground. Then, a sound, which we had heard no sound previous to this, the sound of, like of, ah, you've heard diesel locomotives revving up. (note 4) That's the way this thing sounded. And, then, we were attacked, I guess you would call it. In any event, we were swept by some form of a ray that was emitted in pulses, in waves that you could visually see only when it was aiming directly at you. That is to say, like a searchlight sweeps around and the segments of light you would see it coming at you.

"Now you would feel a burning, tingling sensation all over your body, as though something were penetrating you. And ah, so the company commander, Lt. Evans, hauled us into our bunkers. We didn't know what was going to happen. We were scared. We did this. These are underground dugouts where you have peep holes to look out to fire at the enemy. So, I'm in my bunker with another man. We're peeping out at this thing. It hovered over us for a while, lit up the whole area with its light that I'm telling you about, and then I saw it shoot off at a 45 degree angle, it's that quick, just, it was there and was gone. That quick. And it was as though that was the end of it. But, three days later the entire company of men had to be evacuated by ambulance. They had to cut roads in there and haul them out, they were too weak to walk. And they had dysentery and then subsequently, ah, when the doctors did see them, ah, they had an extremely high white blood cell count which the doctors could not account for.

"Now let me inform you on this. In the military, especially the Army, each day you file a report, a company report. Now, we had a confab about that. What do we do about this. Do we file it in the report or not? And the consensus was "no." Because they'd lock every one of us up, and think we were crazy. At that time no such thing as UFO had ever been heard of and we didn't know what it was. And I still don't know what it was. But I do know that since that time I have periods of disorientation, memory loss, and ah, I dropped from 180 pounds to 138 pounds after I got back to this country. And I've had great difficulty keeping my weight up. Indeed, I'm retired and disabled today."

The following questions were asked the witness by Mr.Timmerman immediately following the above narrative.

Q. "What has been your employment since you were in the military?

A. "I was with the American tobacco company."

Q. "In what capacity?"

A. "Manufacturing."

Q. "In a plant near here?"

A. "Yes, In Reedsville." (North Carolina)

Q. "I see. How long did you work in that plant?"

A. "Twenty six and one-half years."

Q. "How long ago did you retire?"

A. "1969."

Q. "I see. And at present you are retired?"

A. "I'm retired, yes, yes."

Q. "OK. ... Do you have the names of any of the men in that unit?
Do you recall were any of them from this part of
the country?"

A. "No, no."

Q. "They were from all over?"

A. "Uh huh (yes). They were from California ..."

Q. "Did you maintain contact with any of them?"

A. "No, we've lost contact with them...so many of them got
killed off after that."

Q. "Well, it was 36 years ago too. Did you ever before now
have occasion to discuss this case with anyone else."

A. "Yes. I've told my wife and my family - my wife and
my children. I have related this story to them many
times over, since comin' back from Korea. But,
you know how it is with things like that. They
said, "uh huh," and they'd go on, and that's the end of it."

Q. "Right. Well, this case, ah, there's the possibility
that we may have other cases, other reports in the file..."

A. "Oh, there's one thing I forgot. I'm not tryin' to add to the
story but there is one thing that's important. You know I

told you I fired at it with the M-1 rifle... made contact. And the thing went wild like it was gonna' fall."

Q. "Right."

A. "Alright, subsequently, we opened up with everything we had and after that nothing would affect it...That one shot got it. But evidently, their defences were lowered, briefly, and when I connected, and when their defences were thrown back up, and after that, nothin' could hit it."

Q. "Did the bullets seem to just move right through it?"

A. "Well, they..."

Q. "There was no contact?"

A. "No contact. But the first time I did connect. And it was metallic because it was an armor piercing projectile from an M-1 rifle. And we did hear the metal to metal, as it impacted."

Q. "Was it almost instantaneously, because apparently the object wasn't very far away."

A. "It wasn't far away, no. It was hovering right above us, like that, about like that ah, ceiling there...and apparently observing us." (Note 5)

Q. "Yeh, I see."

A. "It apparently had no hostile intent at that time, until I fired."

Q. "Did you recall whether there was any sound associated with the event?"

A. "There was no sound until I hit the object. That's what I'm tellin' you, that this sound of locomotives revving up, diesel locomotives. Yeh, a deep sound. (cf. note 4) And that thing was winding up...and it, it recovered from this impact. And it was fully operational."

Q. "So that the projectile you fired apparently didn't cause any serious..."

A. "No, it recovered from it. Ah,...but I did see the blue-green light on one of your pictures there...like I described to you."

Q. "Did you...? The light that you saw, was in this...(photo exhibit)."

A. "It was in one of the pictures here. It was orange in the beginning and then it changed to a blue-green light."

Q. "I see."

A. "I do believe that these things are real and I think that there is a cover up, and we were ordered to say nothing about this. That shows you they are covering up.... It is foolish to believe that we have the only technology anywhere, you know? There are other intelligences.... Well I hope, I wish...if I can raise up any of the names of the men, possibly, that are still alive, and I doubt if you could get them to come forth, but if I could, there would be some way to verify this."

"I'm sixty years old now, but back when I was younger there were three days I still can't account for."

Q. "Following the event?"

A. "Uh huh (yes), there's three days I still can't account for. Why, ever since I came back home. My wife can tell you about it but I still don't remember." (note 6)

Q. "Days here at home?"

A. "Uh huh (yes)."

Q. "And you had a period..."

A. "I'd have these extreme headaches. They'd have to send me home, put me in the hospital and so forth."

Q. "And they couldn't explain what was going on?"

A. "Uh uh (no)."

Q. "That's part of your medical history that Dr. R. would have?"

A. "Uh huh (yes).... but you could talk to my wife, ah, what's left of my family, half of them are in the grave... and they'll tell you that I've told the same story many times. And if I was lyin' I think I'd get mixed up in it after a while."

I contacted Mr. Wall in May of 1989 to ask a number of other questions. Here are my questions and his answers.

Q. "Please try to estimate how long a period of time went by from the time your M-1 rifle fired to the time you heard the round strike the object?"

A. "One to two seconds (best as I can remember)."

Q. "Did you ever have severe headaches from the time the object departed to one week later?"

A. "Yes."

Q. "Please try to describe when they first started."

A. "The next day or so after contact with the object."

Q. "What other physical ailments did you experience within the first week after the experience?"

A. "Dysentery, nausea."

Q. "Within the first week or so did you ever vomit?"

A. "Yes...for several days."

Q. "Within the first month or so did your appetite change?"

A. "Yes... I lost my appetite."

Q. "Within the first month or so did your thirst change?"

A. "Yes...my thirst increased."

Q. "How many of the men in your company did you talk to about this strange event at the time?"

A. "Approximately 25 or more."

Q. "Did any of the other men you talked to tell you things that were different from your experience?"

A. "No."

Q. "Have you ever seen such an object before this time in Korea?"

A. "No."

Q. "Have you ever seen a similar object after this sighting?"

A. "No."

Q. "What do you think the object was?"

A. "An alien spacecraft - nothing like I had ever seen."

Q. "What compass direction (approximately) were your guns firing from the hillside toward the town?"

A. "North."

Q. "What compass direction did you first notice the object?"

A. "East."

Q. "What compass direction did the object finally disappear?"

A. "Forty five degrees up (and) then West."

Q. "When were you discharged from the Army?"

A. "June 1952."

Q. "What type of military discharge did you receive?"

A. "Honorable."

Notes:

1. Mr. Wall's name, rank, serial number, and other information was found on the official Army roster for his unit as of July 1951.

2. Historical accounts by Marshall (1951), Michaelis and Davidson (1951), Pearl (1963), and Pullen (1954) provide interesting background information on the 27th Infantry Regiment "Wolfhounds".

3. Lt. Evans' name and other information for him was found on the official Army roster for "E" Company.

4. A very similar event took place on March 17, 1981 on the Columbia River, just east of Portland, Oregon. Several eye witnesses described an orange spherical light which hung above the river while emitting a low throbbing, diesel engine sound that was tape recorded. An analysis of this tape by the *J. Allen Hynek Center for UFO Studies*, Chicago, Illinois could not identify it.

5. The ceiling in the shopping mall where this interview took place was approximately 25 to 30 feet above the floor. Even for a very low muzzle velocity (say 900 fps), his estimate of 1 to 2 seconds duration from when his M-1 fired to when he heard the projectile strike the object is clearly too long. Is it possible that he only heard the echo of his own rifle firing from the surface of the object?

6. It is unlikely that the light ray that was emitted from this object was the cause of his (much later) symptoms of his memory loss.

In another letter to Mr. Wall dated April 14, 1990 I provided him with a list of 72 names of personnel who were in his company with the request that he try to recall (and mark on the letter) the names of as many of them as possible. His reply of June 1st was terse; he wrote the word "none".

General Comments:

This close encounter of the first *and* second kind contains interesting corroborative data to the March 10, 1951 pilot sighting near Chinnampo, North Korea, the January 29, 1952 sighting by three military pilots flying near

Wonson at night, the February 24, 1952 bomber crew sighting over Antung, and the May 31, 1952 pilot sighting south of Chorwon (all presented in the next chapter). This particular sighting report also contains valuable details related to the nature of this aerial reality such as apparent three-dimensionality and mass, emitted noises and luminous radiation in a partially collimated fashion (in seeming reaction to the impact of the soldier's rifle projectile), biological interaction (immediate skin sensations and subsequent skin surface and internal sequelae), and a responsiveness that is (allegedly) in direct response to the witness' "aggressive" behavior. The identity of what is behind this phenomenon remains shrouded in mystery.

A number of issues of Jane's All the World's Aircraft around the year 1950 were studied with regard to the kinds of American and Russian aerial devices that were in existence. There was no mention of any type of remotely piloted vehicles (RPV) nor other kinds of non-aerodynamic vehicles.

Chapter 3

American Pilots Report UFOs over Korea

This chapter presents 30 UFO sighting reports that were made by U.S. Air Force, Navy, and Marine pilots during the war. They are interesting because they show that the UFO was capable of out performing the combat aircraft the U.S. had at that time.

Sept. 1950 **Korea**

A U.S. Navy plane on a combat mission was approached by two large discs. The aircraft's radar was jammed, and its radio transmitter was blocked by a buzzing noise each time a new frequency was tried. (Hall, pg. 73, 1964). A more complete report provides the following information (UFOIRC, 1966).

> "Very early in the morning, three fighter-bombers took off from the flight deck of a U.S. aircraft carrier. The sun hadn't risen and there was bite in the air. Two men -- a pilot and a radar gunner-- occupied each of three planes assigned to a routine mission, bombing and strafing a truck convoy that combat intelligence expected to be winding along the floor of a valley about a hundred miles from the Yalu river. The takeoff was routine, as was the flight to the target area. At about 7 A.M., just as the sun was breaking above the mountains in the East, the aircraft were proceeding North, ten thousand feet above the valley floor. Radar observers had their eyes peeled for the target.
>
> "I was watching the ground below for the convoy, reported ... and was startled to see two large circular

shadows coming along the ground from the Northwest at a high rate of speed. We were flying North above a valley which was surrounded on the East and West by mountains, with a pass directly ahead of us to the North. When I saw the shadows I looked up and saw the objects which were causing them. They were huge. I knew that as soon as I looked at my radar screen. They were also going at a good clip -- about 1000 or 1200 miles per hour. My radar display indicated one and a half miles between the objects and our planes when the objects suddenly seemed to halt, back up and begin a jittering, or fibrillating motion. My first reaction, of course, was to shoot. I readied my guns, which automatically readied the gun cameras. When I readied the guns, however, the radar went haywire (note 1). The screen bloomed and became very bright. I tried to reduce the brightness by turning down the sensitivity, but this had no effect. I realized my radar had been jammed and was useless. I then called the carrier, using the code name. I said the code name twice, and my receiver was out--blocked by a strange buzzing noise. I tried two other frequencies, but couldn't get through. Each time I switched frequencies the band was clear for a moment, then the buzzing began. (note 2)

"While this was going on the objects were still jittering out there ahead of us, maintaining our speed. About the time I gave up trying to radio the carrier the things began maneuvering around our planes, circling above and below. I got a good look at them. I had never seen anything like them before, and I learned after we reached our carrier that the other men in that flight had the same opinion. They were huge. Before my radar set was put out of commission, I used the indicated range plus points of reference on the canopy to determine their size. They were at least 600 or possibly 700 feet in diameter.

"The objects had a silvered mirror appearance, with a reddish glow surrounding them. They were shaped somewhat like a coolie's hat, with oblong ports from which emanated a copper-green colored light which grad-

> ually shifted to pale pastel-colored lights and back to the copper-green again. Above the ports was a shimmering red ring which encircled the top portion.
>
> "When the things maneuvered above us, we saw the bottoms of them. In the middle of the underside was a circular area, coal black and non-reflective. It was simply inky black, and it is important to note that although the whole object jittered while maneuvering, the black circular portion on the bottom was steady and showed no indication of movement. (note 3)
>
> "When the objects seemingly finished their inspection of the Navy planes, they took off in the same direction from which they had come, and disappeared at a high rate of speed." (Ibid., pp. 3-4 and 3-5)

Notes:

1. The apparent reactivity of the UFO phenomenon to the pilot's volitional act is a commonly reported event in UFO literature.

2. The short delay which occurred suggests a very rapid, automated means of (a) scanning all emitted frequencies from the aircraft, (b) locating the new radio frequency being used, and (c) activating an interference signal at that new frequency.

3. There is a close similarity between this description and that given by a commercial airline flight crew over Lake Michigan on July 4, 1981. See (Haines, 1982; 1983). Another case which the author has studied includes similar visual characteristics. An unidentified circular, apparently metallic object having a dark spot centered on its upper surface was photographed in 1971 by an official Costa Rican government airplane during a ground mapping flight (Haines and Vallee, 1989; 1990). Our extensive photographic analysis could not provide a reasonable explanation for the image.

March 10, 1951 0051 GMT Chinnampo

The chief eye witness in this case was the left scanner on a B-29 (#5369) flying at almost 17,000 feet altitude on a heading of 30 degrees, 42 miles SW of Chinnampo, Korea at 0051 hrs GMT (note). Eight other crewmen saw very brief glimpses of this bright light. The night was clear and moonless. The main witness saw a flash of brilliant light at his 9 o'clock position that moved slightly downward and to the rear of the airplane. The scanner "... reported it as a flare." The phenomenon appeared as a "reddish-yellow glow that dropped slightly to (the) same elevation of aircraft then burst with bluish-white brillance. It was estimated to be as large as a basketball and did not deviate from (a) straight course toward aircraft or curve right or left. Tail gunner attracted by brilliant light and saw flash over (his) left shoulder off right wing. (He) did not observe any movement or any color other than silvery bright flash. Bombadier saw bluish-green flash from (his) side vision. By (the) time he could turn his head it was gone.... Pilot saw object at 3 o'clock from corner of (his) eye as a flash of blue-white or blue-green. None able to describe red-orange tail. None saw any shape or shadows or anything resembling exhausts trace or another aircraft.... None could estimate (the) distance on anything except sheer guess." (USAF, Staff Message Division report no. A 5394 INT-IR, dated 26 March 1951).

The Air Force listed this case as UNIDENTIFIED in its Blue Book files with the possibility of it being flak, a flare, or a meteor.

Note: GMT refers to Greenwich Mean Time.

July 1, 1951 2250 Korean Time Seoul

This U.S. Air Force Air Intelligence Information Report IR-33-52 dated 16 July 1952 was prepared by ATIL Office, D/I FEAF's Captain Charles J. Malven (ATLO), 1st. Marine Air Wing. It states:

> "At 2250/K, four night pilots in different areas simultaneously reported a large green ball, very bright, and trailing streaks of red which soon turned blue, passing overhead at 10-15000 feet on North-Northeast heading. Pilot's locations at time of sighting were as follows: One F4U-5N 2 miles North of Seoul, one F4U-5N at CT 1070, one F4U-5N at CT 4020, one F7F-3N at CU 2010. Pilots concur that object was moving with tremendous velocity."

The official Air Force conclusion was "METEOR" due to the visual description given.

Fall 1951 Night Location Unknown

More than 14 U.S. Navy ground and airborne radar sets tracked a UFO which circled over the U.S. fleet. The object moved at speeds ranging from "slow" to over 1,000 mph and was at an altitude of 5,000 feet. The witness, Lt. Cmdr. M.C. Davies, U.S.N. was on board a CVE class aircraft carrier at the time as a Naval pilot (with over 4,000 flight hrs). Following is his report submitted to the National Committee on Aerial Phenomena (NICAP).

> "It was at night, I was riding with a radar operator which I often did to check on their proficiency. We were flying at 5,000 feet, solid instruments, with our wingman flying a radar position about 3 miles astern and slightly to our right or left. The target, which was slightly larger than our wingman, I picked up on our scope, had been circling the fleet; it left the fleet and joined up on us a position behind our wingman, approximately the same position he held on us.
>
> "I reported the target to the ship and was informed that the target was also held on the ship's radars, 14 in number; and for us to get a visual sighting if possible. This was impossible because of the clouds. The target retained his relative position for approximately 5 minutes and then departed in excess of one thousand miles per hour. He departed on a straight course and was observed to the maximum distance of my radar which was two hundred miles.
>
> "Upon completion of my flight an unidentified flying object report was completed, at which time I was informed that the object was held on ship's radars for approximately approximately seven hours." (Hall, pg. 84, 1964).

Note: A weather station was installed by the Air Force at Chip'o Ri, coordinates CT 5222. The largest inflatable weather balloon used there was a 30 gram "ceiling" balloon with a maximum diameter of 15 inches. (See Appendix 1 for balloon characteristics. Note the error in the type of balloon and its diameter.)

Jan 29, 1952 2300 and 2324 Korean Time Wonson & Sunchon

Three crewmembers (tail, left, and top gunner) of a B-29 reported a five-minute encounter with a light orange colored sphere which shot away at an angle. (Newsweek, March 3, 1952; LIFE Magazine, April 7, 1952). This public information came from an Air Force press release. The more complete official Air Force Air Intelligence Information Report IR-2-52 dated 10 February 1952, was prepared by 2nd. Lt. Mario Perez, Intelligence Officer, 98th. Bomb Wing Intelligence Office. It states:

> "1. Observed one globe shaped object with a slight tear drop effect noted on lower side, estimated size to be three feet across when at its nearest point to B-29. The size was not definitely established as the distance from the observer was not known. The color of the object resembled the sun, a light orange, and it occasionally changed to a bluish tint (note 1). The outer edge of the object appeared to be fuzzy and it seemed to have an internal churning movement like flames or fiery gases when it was at its nearest point to the B-29. The object was first observed parallel on course at eight o'clock level, where it seemed to be about the size of a saucer, gradually becoming larger as it approached the B-29. It came in on the same level as the B-29 and remained in the same relative position to the B-29 for approximately one minute and then receded on the same path, fading away in the distance."
>
> "2. Sighted on January 291324Z for a period of approximately one minute. The left Gunner and Tail Gunner observed the object without the use of any optical or electronic equipment. The B-29 was at 22,250 feet, and ground speed (of) 125 knots.
>
> "3. The two observers were airborne in the B-29 in the crew positions of Left Gunner and Tail Gunner. Object was observed at 3926N 12555E and level with the B-29."

Other Air Force files provide additional information. The orange object flew beside the first B-29 for about five minutes during its encounter at 2300 Korean time and beside a second B-29 for one minute at 2324 Korean time. Its location was approximately 39 deg 03 min N and 127 deg E. The military coordinate position was cited as CU-4020 or about 50 miles off the south coast during the first sighting and at YD-5459 or about 5 miles SSE of Sunchon. The second aircraft received flak shortly after sighting the object. The B-29 was flying at an altitude of 22,500 feet on a heading of 274 deg. The light was seen "shortly before receiving flak. The object approached the B-29 from between 8 and 9 o'clock level then withdrew and disappeared at the same clock position." It was said to be a shimmering (assumed to be spinning)" globular" object with a color that resembled the sun with "an occasional bluish tint" and appeared to be about three feet diameter (at its nearest position). All witnesses had had extensive flight combat experience in WW2 and stated "emphatically that the sightings noted above bore no resemblance whatsoever to anything they had previously experienced."

"It is the opinion of this officer that these sightings represent another example of new technique in warfare under test by the enemy. Comment by D/I, FEAF Bomcom; it is assumed that there is still a very real possibility that these phenomena may very well indicate the presence of new enemy flare devices, despite the unit intelligence officer flling (sic) that such possibilities are ruled out. Aside from that, it is worth mentioning that the 98 wing commander was present during one of the subject interrogations, and warned the crew members as to their responsibilities in reporting such observations. Special report of unidentified flying objects W/B submitted. (In) view of utter lack (of) similar observation requires your comments regarding (the) above."

In yet another Air Force report on this case (T52-3047-2, ID. 310956Z) from the Commanding Officer of ATIC, it is learned that the exhaust stack flame from aircraft engine running rich (is) considered doubtful because no known Soviet aircraft have exhaust stack arrangement that would produce this illusion. It was also considered doubtful that the light could have been produced by a jet aircraft engine, afterburner exhaust, ramjet helicopter air-to-air weapon, fireballs, or trailed bombs (also used by Germans in 1942 (WW-2) as a defense by bombers against pursuing flighters. It was a 20-25 kg bomb attached to a wire about 100 to 250 meters long behind the bomber which flew in wide circles and could be exploded at will by a crew member. Such an object might appear to be spinning and to fly parallel with the enemy bomber. Of course in this case there were no enemy airplanes anywhere near the observers' B-29!

Peter A. Stranges, staff member of the Reaction Power Plants Group, Propulsion Branch at the Air Force's Air Materiel Command, Wright-Patterson AFB responded to an information request from Lt. Ruppelt at

Project Blue Book office concerning the possibility that this UFO was a number of conventional objects (viz., jet engine, afterburner, ramjet, pulsejet, rocket engine). His ATIAA memo dated 20 February 1952 stated:

> "2. This group concurs that the objects appear to be propelled but from the descriptions and the time durations cited it is not considered that the objects are conventional jet engine, conventional jet engine with afterburner, pulsejet or rocket propelled. Slight credence, however, is given to the possibility of a ramjet powered helicopter or a modification thereof, with provisions incorporated for exhausting along a section of the trailing edge of the rotor blade, creating the fuzzy edged, internal churning of flames and gases, globe shaped phenomena observed recently by the B-29 crews in Korea."

A memo by Lt. Col. Hundt, AFOIN-V/TC, prepared on 29 February 1952 identified the Soviet fighter aircraft of possible interest as the La-9 or La-11. Both have a cluster of exhaust stacks on each side of the fuselage. There is virtually no correspondence between such an engine exhaust configuration seen at night and the orange globe reported by these witnesses.

In a memorandum prepared by Capt. Fournet for Air Force public relations use he states, "The sightings mentioned, although of a different nature, as is usual, are not abnormal occurrences in the combat theatre. During World War II over both Germany and Japan, combat crews reported sightings of a multitude of these types of objects which could not be identified or explained. These very often were reported as "fireballs," particularly from combat crewmen participating in flights over Japan. *Very often these unidentified objects were reported to have performed violent maneuvers, travel at very high speeds, travel at very low speeds to the point of hovering, followed or flew formation with friendly aircraft, disappeared or disintegrated while in flight, etc. During the whole of World War II and subsequent thereto, intelligence was never able to prove the existence of such unconventional phenomena nor was it able to determine the characteristics of these objects, if indeed they did exist.* A summary of all such related incidents was made essentially as follows. (Italics mine)

Note: The text which follows was (apparently) suggested to Air Force Public Relations for release to interested parties rather than the detailed information contained within the original witness report.

> "The sightings were the result of natural normal occurrences while in combat, e.g., flak, flaming and exploding aircraft, reflections on canopies, searchlight reflections, engine exhaust trails, air-to-air bombs, etc., and sometimes a result of combat fatigue, particularly on the long missions required to bomb Japan."
>
> "3. There have been numerous reports of objects which crews have been unable to identify, while on combat missions over Korea during the present conflict. In general, these sightings have assumed a pattern which parallels the overall pattern of World War II reports. The latest reports from Korea have mentioned the employment by the Communist Forces of an airborne searchlight, possibly attached beneath the fuselage of a conventional aircraft and used in conjunction with antiaircraft and/or fighters. The evaluation of this branch of the sightings outlined in the subject above is that these unidentified objects probably are such airborne searchlights which because of combat fatigue, conditions of sightings, weather factors, etc., were assumed to be objects unattached to anything else. Such a report would appear reasonable in view of the fact that an object of this type would naturally become a center of interest, so to speak, and would tend to cause the observers to concentrate on its details rather than to look for anything else in conjunction with it." (note 2)

An Associated Press wire story dated February 20, 1952 from Tokyo concerning these sightings stated: "Far East Air Force Headquarters in Tokyo, which directs B-29 bomber operations in Korea, would issue no statement on the latest version of flying saucers. Asked if pilots of night fighters or B-26 light bombers had seen the objects, a Fifth Air Force spokesman in Korea said: "To affirm or deny lit would put us in the position of discussing it and we cannot discuss it." (note 3)

Notes:
1. Chapter 2 presents a ground eye witness account of a relatively small round orange object which also changed to blue as seen from a ground artillery

position overlooking Chorwon.

2. This explanation seems to better fit the sighting of February 23, 1952 (below) than it does to this case.

3. Aircrew members were already under a lot of combat related stress. It is likely that one reason USAF officials did not want to openly admit the existence of an actual aerial phenomenon not of enemy origin was to try to help reduce this stress.

February 23, 1952 1415 Korean Time Sinuiju, North Korea

Crewmembers of a B-29 reported sighting a UFO while flying over Sinuiju located near the Yalu River (40.04N; 124.25E). No other information is available on this sighting.

February 24, 1952 2315 Korean Time (Night) Antung

The navigator of a B-29 observed a UFO about 45 seconds after the airplane in which he rode was illuminated by a searchlight. The object was cylindrical in shape and approached from the 7 O'Clock position. It had a "rapidly pulsating gaseous type of exhaust trail... about three times the length of the object with both the tail blast and the object bluish in color." (note 1)

The previously classified Air Force report (IR-1-52; dated 27 February 1952; No. T52-5836) prepared by Captain I. O. Mitchell, Flak Intel. Officer interviewed Captain R.W Smith, 345 B.S. 98th. B.W. the eye witness. Segments of his report follow.

> "At 2315I, a B-29 on leaflet mission at XE 2838 heading 030 deg, altitude 22,000 ft, was picked up in searchlight beam by two lights on right side of acft. Within 45 seconds after acft was illuminated by searchlights, navigator observed object approaching acft from position, on ground estimated to be at XE 2342 (note 2). No launching blast was observed and object appeared to approach acft at an angle of 30 deg from the left rear. When observer initially spotted object, it appeared to be round in shape and bright blue in color. As it came closer to the acft, the shape

appeared cylindrical with tail approximately three times length of the forward mass of object. The color of the tail was the same as object and appeared to be rapidly pulsating gaseous type propulsion blast. Observer saw object for an estimated 45 seconds. The course of the object was at an angle of 45 deg from horizontal, of high velocity, flat in trajectory until it leveled off approximately 4000 ft. from acft. It assumed level flight characteristics, turned to the left as though making a correction to intercept acft at a predicted point. The object continued to close in toward interception point on level with acft for about 15 seconds and then broke off in flight and head down and underneath acft at gradual angle. At the closest point, the object appeared to come within approximately 3000 ft. of acft and appeared to be the size of an automobile. As it broke off from level flight, the tail diminished rapidly in size and became brighter in color. The indicated airspeed of acft was 200 knots and acft remained at 22,000 ft. altitude. Searchlights continued to illuminate acft and meager, inaccurate flak was encountered during entire period object was observed. Searchlights worked in multiples of two and four light pickups, and kept ship illuminated for approximately four minutes. ECM operator received weak electronic signals, but was unable to take electronic countermeasures. Pilot executed evasion action in a series of "S" curves, then made a shallow dive and searchlights and flak ceased. Weather was clear and visibility unlimited."

Notes.

1. These characteristics are similar to those described in chapter 2.

2. It is of interest to observe that the cylindrical object approached the bomber from the direction of the ground *after* the aircraft was illuminated by ground searchlights.

March 29, 1952 Korea

The pilot (Brigham) of an F-84 sighted a UFO over Korea. No further information is available. (Hall, 1964)

May 10, 1952 Korea

Project Blue Book monthly summary for the period 1 - 10 May 1952 includes a single entry for a sighting on May 10th by a military observer that was evaluated as BALLOONS. No other information is available.

May 15, 1952 2000 KoreanTime North Korea

Air Intelligence Information Report 52-79 dated 28 May 1952 describes a brief but interesting encounter with a UFO by two "very reliable" pilots of the 51st Fighter-Interceptor Wing, 25th Fighter-Interceptor Squadron. The two airplanes were on Mission C/51-02 flying wing positions in maple flight. Lt. McCarthy was the first pilot to see the strange object. He had completed 27 combat missions. The other unnamed pilot was the Interceptor Group Operations Officer with 56 combat missions completed. Following is a summary of the key events:

A silvery oval shaped object larger than a MiG jet airplane was seen at the 9 o'clock position below the two F-86E jet airplanes at an estimated altitude of from 8,000 to 10,000 feet and about twenty miles away. The single seat airplanes were on a heading of 280 deg; the object continued on in a straight flight path and disappeared at about the 3 o'clock position. While flying at an altitude of 30,000 feet and an airspeed of 500 knots, both eye witnesses saw the object for only 3 to 5 seconds and noted that the object travelled at an estimated 1,200 to 1,500 mph in a "rolling maneuver" (note). The reported weather was a medium haze with visibility of between 10-12 miles.

Note. The meaning of this is unclear.

May 15, 1952 1835 Korean Time North Korea

Air Intelligence Information Report No. 52-81 describes an encounter between the pilot of an F-51 propellor-driven fighter bomber and a silver object estimated to be about 50 feet in diameter. The aircraft was assigned to the 18th. Fighter Bomber Group. The pilot had flown 76 combat missions in WW2. He was at military coordinate CT 8000 flying on a heading of 180 deg at an altitude of 9,000 feet; he first noticed the object at the 1:00 o'clock position. "By the time I called the object in as a bogie, it had moved to the 3 o'clock position and started a steep climb, which I assumed was the beginning of a loop, but at the vertical position in the loop, the object moved forward in its original course of travel (360 deg) and wavered momentarily and then descended and disappeared into the haze which reached an altitude of approximately 7000-8000 feet." (Ibid., pg. 3) While the airplane was travelling at only 240 mph, the UFO was estimated to be moving at 1,000 mph first to the east and then changing to the north. It was at approximately the same altitude as the airplane during the 15 - 25 second-long sighting.

May 26, 1952 0320 Korean Time North Korea

This Project Blue Book case (Air Intelligence Information Report No. 52-85) was classifed as a POSSIBLE BALLOON on the record card. Both crewmen on the F-94 jet interceptor (Figure 7) saw the brilliant white object (the pilot for about 14 seconds) in addition to positive ground and on-board radar contact with the same object. The weather was very clear. The signed pilot sighting report is perhaps the clearest description of what happened.

> "While on CAP-sortie from K-13 Bromide informed us of an unknown located at CT 4856 and told us to investigate. 'After descending to 2500 feet we saw, in the glare of the front line searchlights, a small plane beneath, but were unable to descend farther due to terrain to determine type. Bromide told us to drop it and take up a southerly heading. When we turned to this heading, we saw a brilliant object above and in front of us and asked Bromide if we could investigate. With his permission we made our first pass from north to south within 600 feet, then a succession of passes from cardinal compass points ranging to within 50 feet, but, because of the brilliancy of the light, we could discern no outline. On our last pass my Radar Observer looked back and said he thought he saw an object on our tail. I did not give much credence to his observation until Bromide called and said he painted an object on my tail. I then requested Bromide to vector me for an interception on this new unknown. (At this time the object was still

Figure 7

F-94C Starfire Jet Interceptor
(from author's collection)

in sight.) When I received my first vector I was south of the brilliant object and started a starboard turn to the intercept vector for the new unidentified. On a northeasterly heading my R/O gave me a lock-on, overtaking 50 knots, range 7000 yards, nearly dead ahead and slightly above. When I closed to 6000 yards, the object started to pull away and I threw the afterburner in (airspeed at this time was 250 knots). In a gradual climb and still with the object nearly centered on the scope, the range increased at a steady rate from 6000 yards to approximately 26000 yards in 12-15 seconds. At that range we broke lock. No evasive action was observed and no exhaust patterns such as might be expected.

Signed: 1st Lt. Martin -------

The Radar Observer provided the following signed statement:

> "We had dropped down to about 3,000 feet to investigate an unknown that GCI was painting, and after making a few circles over this unknown, Martin, my pilot, spotted this bright light. We got permission to investigate the light by GCI. I tried to pick the light up with my radar, but I could not. GCI said they were then painting the light. We made about seven (7) passes at the light; after the third pass I came out of the set to take a look. It appeared to be a bright blue light. I can not understand how GCI was painting this target and I could not pick it up. We were at about 7,000 feet and then after the last pass GCI said that someone was on our tail. I turned and thought I saw a plane, but could not be sure, as it was only for a second. Martin started a right turn, and to climb. I then picked up our target at about 7,000 yds and about 7,000 feet altitude and locked on. I believe that this was the target GCI was painting all the time, and that he was in our area while we were making passes at the light. We started to overtake our target and he moved in from 7,000 to 6,000 yds. He was in the center of my scope, and about 5 or 10 degrees high. We were climbing all the time. He then went from 6,000 yds to about 24,000 yds in about 14 seconds and we broke lock. We did not pick him up the rest of our mission."

Signed: 1st Lt. E ------

The official report points out several other facts of interest which were elicited by the Director of Intelligence, FEAF: (1) The airborne radar set type was an APG-33 and was in very good working order *before and after* the flight in question, (2) The R/O did not take his eyes off the scope at any time until the lock was broken at 24,000 to 26,000 yds, (3) the pip was centered and about 10 deg high at all times until lock-on was broken, (4) the object's acceleration was constant (it did not jump), "The unidentified went into a shallow climb with the F-94 following up to and passing 10,000 feet altitude until lock-on was broken, and (5) since the radar set was on long range it wasn't possible to determine the size of the object. (italics mine)

The Deputy for Intelligence, FEAF wrote:

> "1. The distance covered by the object is approximately 68,500 feet. Acceleration is 650 ft/sec/sec, (or approximately 20 "G" acceleration, (which is much more than the human body can stand for more than a fraction of a second). The final velocity is found to be 9450 ft/sec, or approximately 5580 knots (6420 mph).
>
> "2. The radar sets are checked before and after every mission, and in this case appeared to be working normally. Even if it were assumed that the operator's timing, or the APG-33 calibration were off as much as 50%, the speed is still above any known type of missile.
>
> "3. No information has yet been received from the ground controller at Bromide as to the returns plotted or the length of time that object was held in contact by the ground radar.
>
> "4. The phenomenal acceleration reported leads to the belief that the APG-33 was not operating correctly at the time of contact, despite checks before and after the mission."

Signed: Charles Y. Banfill
Brigadier General, USAF
Deputy for Intelligence

The conclusion of the Deputy for Intelligence stated above illustrates once again the lengths that people will go to to ignore obvious facts that do not agree with their presuppositions. The General conveniently ignored the fact that both aircrewmen visually sighted the same aerial object that the two radar sets had "captured." During times of war it is easier to blame a radar set as being the cause of a UFO report than to have to cope with ever-increasing paperwork and ridicule from "above" for admitting that these two aircrew actually saw what they said they saw.

On the last page of this file is a typed summary sheet with the words "Information Only - Solution: Possible malfunction of airborne radar set,

Source: Military". Nevertheless, both flight crew saw a "brilliant white light straight ahead" right where their radar, and ground radar said it was! In addition, their radar system had been found to be in good working order both before and after this flight making this solution both unacceptable and ridiculous.

May 30, 1952 1900 Korean Time Sea of Japan

The following report was found in Air Intelligence Information Report IR-4-52 dated 7 June 1952 regarding a sighting of an unusually shaped aerial phenomenon by three aircrew of a C-54 aircraft (Figure 8) flying at about 34 deg 15 min N; 139 deg 30 min E over the Sea of Japan.

Following is the signed report by the pilot, Lt. Robert J. Arblaster, AO 1908891.

"On 30th of May 1952, climbing to 3,500 feet, enroute to Oshima Range, on a heading of 190 degrees, at approximately 1925 hours, the co-pilot of my crew remarked that he saw an odd shape in the sky at about 4 o'clock.

"I was flying the aircraft and raised up out of the left seat to get a look at what the co-pilot referred to. To me, it looked like nothing more than a dark cloud, round and towering. It was approximately 60 to 70 miles from us reaching to about 15,000 feet.

"Since I was satisfied that is (sic) was only another cloud formation, I did not deem it necessary to contact a GCI station and busied myself with flying duties. I have enclosed a sketch of phenomena observed.

Signed: Robert J. Arblaster

The co-pilot, Captain William Leet prepared the following account of what he saw.

"At 1855, 30 May 1952, we took off from Tachikawa Air Base on a Troop Carrier trip in C-54 #2452. I was co-pilot on the crew. We climbed on course to Oshima to 8,500 feet, on a magnetic heading of about 180 degree and at a rate of about 500 feet per minute. Our indicated airspeed was 150 - 160 mph.

"From time to time after take-off and after climbing a few minutes I glanced toward the west observing the effect of the sunset behind the cloud formations. I noticed what at first appeared to be a round black cloud detached from the rest of the formation and in a sort of trough in the contour of the tops of the clouds. After watching this for about one minute, I noted that my watch read 1908 hours, as I decided by this time that the object was not a cloud nor did it seem to be any type of aircraft or helicopter.

"I called the attention of the aircraft commander, the engineer and navigator, who was not on the crew but was in the navigator's compartment, to the object. The aircraft commander, 1st. Lt Arblaster, was occupied piloting the plane and glanced only once at the object. The engineer looked at the object for a minute or two and remarked that it was not an aircraft nor a cloud in his opinion. The navigator told me later when I questioned him, that he merely glanced at the object and thought it was a cloud.

"I observed the object until it disappeared from sight at 1914. It appeared darker than the clouds and an almost circular shape but somewhat flattened on top and bottom. At some moments it seemed to wobble slightly. It hovered perfectly motionless from the time I first saw it at 1907 hours until 1913, when it started moving to the west on a heading I estimated to be about 285 degrees. It moved very rapidly and was out of sight at 1914.

"It seemed to move on a track away from our position and horizontally. I watched it until it became a mere speck and disappeared from view.

"I noticed another C-54 during the time I was watching the object. It was at a higher altitude than ours, about 10,000 feet and was to the northwest of our position headed in an easterly direction.

"The sky to the west was pink and the sun was not glaring. I did not observe any exhaust, flame or vapor trails around the object at any time. I am positive that the object was not

Figure 8

C-54 in Flight
(Reproduced by permission of the
National Air and Space Museum
Smithsonian Institution)

an aircraft of any type I have ever seen. It was not a balloon and it was not a cloud.

"I could only guess at the altitude and distance of the object. I estimate it was about 20-25 miles west of my position and at 10,000-15,000 feet altitude.

Signed: William D. Leet

Another signed eye witness report prepared by Sgt. Andrew J. Roe, flight engineer on board the airplane, agrees in most respects with the above two reports and is not included here.

The official explanation for this sighting was "CLOUD." This conclusion did not take into account the clear and specific description provided by the co-pilot that: (1) none of the observers thought it was an aircraft, (2) it was a circle, flattened on the top and bottom, and (3) it appeared to accelerate away from the observing airplane in less than a minute.

May 31, 1952 0420 Korean Time **South of Chorwon 38.02N 127.07E**

In this USAF pilot sighting the official conclusion was either an "AIRCRAFT" or a "BALLOON WITH A FLARE." Details are from Air Intelligence Information Report No. 52-106 obtained from an F-94 flighter pilot of the 319th Fighter Interceptor Squadron. While the F-94 airplane's radar achieved a positive lock-on the object ground radar failed to "paint" it. A round, brilliant, bluish-white light was reported to possess "...a superior speed and climbing ability over the F-94 and had the ability to turn as well as the F-94. This object was first sighted visually by ground sources whose statements are included in reports 52-93 and 52-94." The airplane was at 30,000 feet altitude and 450 knots when the UFO was seen. The pilot's report stated:

> "The F-94 descended in a port turn to intercept unidentified object 6000 feet below on a 90 deg course and altitude of 8000 feet. Unidentified object began a port climb at the same time to intercept the descending F-94 and accomplished a maneuver which silhouetted the F-94 against the light of dawn. The F-94 turned on afterburner and tried two quartering head on passes with the unidentified object resulting in neither being able to get astern of the other. Hufberry's maneuvers ensued to 3,000 feet where more passes were exchanged for a few minutes. The unidentified object then increased his speed to an estimated 400 knots on a 45 deg heading and began pulling away from the F-94. When last seen the unidentified object had seemingly increased his speed to approximately 450 knots whereupon the F-94 gave up pursuit at 0355 and returned to base."

Comment: The physical proximity to Chorwon and the similarity of the shape of this object with that described in Chapter 2 raise the possibility that the same aerial object could have been seen in both instances.

June 1, 1952 **Korea**

This case is listed in the Project Blue Book files as an ASTRO (FIREBALL), however, the complete file was gone and listed as "missing."

June 5, 1952 **North Korea**

This military sighting in Project Blue Book files is listed as a "BALLOON" sighting. No information was found, however, other than the summary card.

June 6, 1952 **Kimpo Air Base, Korea**

This military sighting in Project Blue Book files is listed as "UNIDENTIFIED." However, the complete file is listed as "missing."

June 7, 1952 **Seoul**

This Project Blue Book case was listed as "BALLOON" on the summary card. No other information is available.

June 20, 1952 1505 Korean Time **CT 0909 Grid area Korea**

This sighting was found in Air Intelligence Information Report No. 30-52 dated 16 July 1952. The summary sheet signed by Brigadier General Charles Y. Banfill, Director of Intelligence cross-referenced the following previous reports: IR-23-52, D/I FEAF; IR-52-106, and IR-52-85. Following are the salient details from this report:

> "1. Object was oval in shape, from ten (10) to twenty (20) feet in diameter, was either white or silver. There was only one (1) object and no vapor trails or gas trails of any type were noticed. The object approached from the south, made a complete left-hand orbit of the target area just worked by the flight and headed away in an easterly direction, where it was soon lost from sight.
>
> "2. It was first observed at 1503/I and was out of

sight at 1504/I on 20 June 1952.

"3. Visual observation was made from approximately 5300 feet, aircraft were in a left orbit south of their target indicating 180 knots. Type aircraft flown were F4U-4B (Figure 9).

"4. Observers were (at) CT 0909 over the west central sector of Korea. Object was approximately 4000 feet below and about two (2) miles away from the reporting aircraft. Altitude of object was estimated at 1000 feet. It approached from the south and made a 360 deg

Figure 9

F4U-4B Aircraft in Flight
(Reproduced by permission of the National Air and Space Museum Smithsonian Institution)

orbit of the CT 091129 area and headed off on an easterly direction and was out of sight at CT 120118. One (1) pilot made a dive at the object but was outdistanced. Aircraft's speed in dive was 250 knots indicated.

"5. Four (4) pilots in a flight of five (5) observed the object after it had been called to their attention by the flight leader. All of these pilots are well trained and have flown an average of 44 combat missions in Korea. The reliability of the report is considered A-1. Pilots names are:

Captain Bobbie Foster 024474/7302 USMCR
Captain Richard Francisco 038989/7302 USMCR
Captain Teddy L. Pittman 036251/7302 USMCR
Captain Ronnie A. McDonald 039833/7302 USMCR"

June 23, 1952 0608 Korean Time CV 4 Grid Area Korea

A pilot of the 18th Fighter Bomb Group reported seeing a black disc shaped object about seven times wider than thick. "The object was observed at 12,000 ft. altitude descending rather irregularly at 6,000 ft., flew at that altitude for six or seven seconds, approached the observer within 1500 ft., and then disappeared into the clouds. The size of the object was estimated to be 15 to 20 ft. in diameter. The speed was estimated to be approximately 250 mph. This Air Force, D/I Comment: "To date there has been no additional info obtained regarding these sightings; therefore, positive identification cannot be made at this time."

"D/I FEAF Comment: No logical explanation can be definitely given to these sightings (reference to the June 6th case). (This case) could possibly be due to light reflections on the clouds, so the report does not indicate whether the clouds were above or below the observed aircraft."

Note. The D/I's reference to the "observed aircraft" seems to make it easier to simply dismiss this sighting as a misidentification of a manmade object rather than a UFO.

August 9, 1952 2057 Korean Time K-3 Area of Korea

This case was filed as Air Intelligence Information Report No. 37-52 dated 25 August 1952 from the 1st. Marine Aircraft Wing, Korea. Cited as a cross reference was IR-35-52. The file contained a report by a pilot flying at 10,500 feet altitude near coordinate K-3 and by a ground radar observer. First the pilot's report:

> "Richmond 18 (flight name) reports strange non-conventional aircraft passed him 500 yards on starboard wing about 1000 feet higher. Aircraft had ball of fire at the rear with a very long streamer of flame. Object (was) glimpsed at front before passing was not of any type known. Aircraft was going at a very high rate of speed. He notified me when he checked in over K-3 at 2057. I verified seeing several radar blips on scope in direction 205 degrees at approximately the same time Richmond 18 reported seeing the strange aircraft. Very conservative estimate of speed of 600 to 800 knots because speed of 1200 to 1400 knots was too high for one aircraft. It was probably two, one behind the other."

A statement made by 2nd. Lt. Leo N. Nagrodsky 053011/6709 Marine Corp to Commanding Officer, Marine Ground Control Intercept Squadron dated 11 August 1952 included the following ground radar information on this sighting:

> "1. On 9 August 1952 I had the duty as Senior Watch Controller at MGCIS-3 CIC from 2000 to 2400.
>
> "2. On coming on watch I noted that AC-CI phone circuit was weak and instructed the duty telephone man to check it. He advised Devestate Able that their transmitter batteries were apparently weak. Devestate Able stated that they would work on the circuit.
>
> 3. (6 lines of initial text omitted here by author) "On the third sweep Richmond 18 was turning toward K-3 and was about 17 1/2 miles from MCCIS-3. A target appeared at 205 degrees, 27 miles and I started to adjust the scope to get a maximum pick up on that target. On the fourth sweep Richmond 18 was at about 215 degrees, 15 miles, and a target appeared at 205 degrees, 32 miles. I did not connect this target with the third sweep target, because it was so far

from it and thought that two Richmond aircraft were coming up from the south. On the fifth sweep, fifty seconds after the first faint target had appeared, Richmond 18 was at about 220 degrees, 15 miles and a strong peculiar blip about 2 miles in length and curved at an angle of about 80 degrees from the trace of a normal target appeared at 206 degrees, 38 miles. I could not tell whether it was an aircraft or some type of interference and did not connect it with the previous blips. No more unusual blips appeared and I decided that they had all been due to interference of some type. The above bearings and distances are as I recalled them about seven or eight minutes later, but I believe them to be accurate within two miles and five degrees. The radar teller's scope was not peaked as high as mine and he did not see any blips other than Richmond 18. I did not call the radar teller's attention to the unusual blips I had seen, because he was accurately reporting all other tracks, and I believed that interference had caused the widely spaced and peculiar blips.

"4. At 2057 Richmond 18 reported in and then asked if I had seen any plots in his vicinity seven to nine minutes before. I replied that I had seen what would have been either several planes or noise at the time. Richmond 18's comment was that it was not like any airplane he had ever seen and described the object as follows: "a large dark object with (text unreadable) and very long streamers or flare which did not resemble a jet exhaust.

"He estimated that it had passed him about 500 yards to his right and about 1,000 feet above his altitude 9400 feet."

In his assessment of these two reports, Captain Charles J. Malven, USAF, ATLO stated:

"2. Further, the charts admit to an error of 2 miles and five degrees. Therefore the constant course heading of a meteorite seems to be substantiated.

"3. A meteorite entering the earth's atmosphere at 18,000 m.p.h. especially if flying a near-horizontal course, would probably decelerate very rapidly. A speed of 1500 m.p.h. at 10,000 ft. altitude may be concomitant with these conditions."

August 23, 1952 0104 Korean Time Sinuiju, North Korea

This UFO report was classified as a possible MISSILE in Air Force Report IR-3-52 made by a B-29 combat crew of the 19th Bombardment Group, Medium while flying a weather reconnaissance mission in the Sinuiju area (coordinate KE 2841). The report states:

> "At 0104I on 23 Aug 52 over KE 2841, aircraft 5369 which was scheduled for weather reconnaissance in the Sinuiju area, sighted an unidentified object. The aircraft was on a heading of 044 degrees at an altitude of 22,000 feet, weather was 1/8 to 3/8 undercast with no moon. The sighting was made by the Airborne Commander. Aircraft Commander, Pilot ---- C.F.C. and Right Scanner. The object was described as appearing to be 6 or 7 feet long, cigar shaped, orange-red in color and red particles ------ 1 ft. --- ------- appearing to shoot off at right and left from the path of --- -------- trail was seen but the whole object seemed to glow with an orange-red color. The object was first seen at the 3 o'clock level position approximately 300 feet away. It was travelling up at an angle of 45 degrees approximately ----- ----- ------ ----- the B-29 and disappeared when it was 150 feet above the -------- altitude. The sighting lasted not more than 3 seconds after it was first seen. Speed of object was estimated as over 300 MPH. The experience level of this crew --- it missiles and all concur on the above sighting. No other information available."
>
> Signed: Eugene L. Rybicki
> 2nd Lt. USAF
> Asst. Intelligence Officer

Captain Charles M. Clark, Jr., the approving officer, inserted the following remark and his initials at the end of the above report: "This object is believed to have been a rocket propelled ground-to-air missile."

September 16, 1952 **Grid TF-77 Area Korea**

C.B. Scott Jones, the pilot of a Navy F9F-2B no. 122583 Panther jet was returning to his aircraft carrier (the USS Princeton; CV-37) with the bright sun at a high angle in the clear blue sky. In a recent letter to the author he wrote "...the only possible UFO sighting I made was at the end of a combat air patrol over TF-77. We had just been recalled to land. As I rolled into a split-S to descend to low altitude from around 30,000 feet, I saw a silver disc directly overhead. I completed my roll and tried to spot it again, but it disappeared or was only a reflection of the sun on the top of the canopy in the first place. I called Passbook Control (the Princeton's Combat Air Control) and reported the sighting. They said that they did not have anything on radar. I was unable to get the same "sun reflection" even though I took up the same direction I had been on when I started to roll into the split-S. It became a non-event. It would have been otherwise if aircrews had been briefed about the other "UFO" type sightings that were being reported at the time. We received an intelligence briefing before each mission, and UFOs were never mentioned."

Jones went on to say, "What may have been the result of reported UFO activity was that we were briefed on a possible Air Force sighting of a surface-to-air missile firing... (given during) ...the end of the 1952 cruise. I assume now that it was in response to the August 23, 1952 UFO sighting that was assessed as a possible SAM (surface-to-air) firing. My last combat flight off the USS Princeton was on October 15 when we flew in support of what turned out to be an invasion feint off Kojo, North Korea." The witness also provided a photocopy of his war diary for this date. It states, "CAP (during a combat air patrol) - spotted silver boggie (sic) over the force just as we were relieved. No radar contact of bogie."

Signed: C. B. Scott Jones

October 16, 1952 **Grid DT 221413 Area Korea**

Project Blue Book microfilm files contain a summary sheet for the period 16 - 31 October 1952. Two separate sightings were recorded on this date. One from Korea and the second from North Korea. Both were classified as being BALLOONS.

Following is information concerning Air Intelligence Information Report No. 52-276 dated 20 October 1952. It took place soon after sunrise.

"On the morning of 16 October 1952, Lt. Hooper, Instructor Pilot and Lt. O'Neill, Pilot, were flying a combat mission in a T-6 aircraft (Figure 10) over the eastern sector of the main line of resistance. At 0650I, Lt. Hooper was flying the aircraft on a magnetic heading of 280 degrees, 140 miles per hour indicated air speed, 6,000 feet indicated altitude. Lt. O'Neill was checking the MLR for activity when he observed what he thought was a strange aircraft at 6,000 feet altitude, traveling approximately 200 mph and heading on an easterly direction. When he first observed the object it was at approximately UTV coordinates DT221413 and the T-6 was at DT312362. The object was 15 degrees to the right of his course traveling on a magnetic heading of 100 degrees.

"Lt. O'Neill called the object to the attention of Lt. Hooper who did not observe it until the aircraft and the object came parallel. He then saw it at the 3 o'clock position. Lt. O'Neill stated that as he was off the relative positions of the object for Lt. Hooper; (who was attempting to pick it up) the object appeared to be decelerating. By the time Lt. Hooper observed it, the object had apparently ceased to move and was hovering over DT2942. Both observers report the fact that the object ceased movement, but differ as to the length of time of the hovering. Lt. Hooper estimates the time at 15 seconds while Lt. O'Neill reports the time to be approximately 1 to 2 seconds.

"Lt. Hooper dipped the right wing of the aircraft for a better observation of the object. It began to move, accelerating in an east-north easterly direction and disappeared from sight in the haze as the T-6 ---- ---- in a 120 degree. The speed of the object at this time was estimated by both observers to be approximately 800 mph. Lt. Hooper bases his estimate on the time elapsed in it's covering approximately 15,000 meters.

"Lt. O'Neill stated that at first he thought the object was a jet or liasion type aircraft, but upon further observation it did not appear to have any aerodynamic features. Both observers concurred in the statement that the object was perfectly circular in shape and metalic silver in color with no protrusions of any type. Both observers stated that the object did not appear to have any visible means of propulsion and no vapor trails or smoke were apparent. The estimated size was 25 feet in diameter, by comparison with the 37 foot wing span of a T-6 aircraft at the same distance. No sound from the object was audible at any time.

"The object was under observation for approximately three minutes from 0650I to 0653I with Lt. O'Neill observing it for the full period of time. Lt. O'Neill who made his observation with the naked eye, described the object as looking slightly (sic) while Lt. Hooper, who wore Government Issued sunglasses, did not observe anythin other than that it looked just like a "ball bearing."

"The nearest observation was made at a distance of 3 to 4 miles.

"The object was seen against three backgrounds, clouds at the first sighting, hills at the hovering position, and sky as it travelled away, and seemed constant as to size, shape and color to both observers.

"The pilots stated that at no time during their observation did the object increase or decrease in altitude."

Brigadier General Charles Y. Banfill, Deputy for Intelligence, USAF FEAF commented: "Since enemy balloons are known to have been used in this vicinity, this observation was probably of the same location reported in IR-52-124, Det #1, 6004th AISS, dated 24 June 1952.

In Lt. Thomas A. O'Neill's official statement made on 17 October 1952, he described the sphere as "...25 feet in diameter and silver in color, and seemed to glow." He went on to say, "Object then changed path of flight to the East by Northeast heading, accelerating to a speed of approximately 800 MPH...the object flew out to sea out of our sight."

This is a very high speed balloon indeed!

Figure 10

T-6 Aircraft in Ground Bunkers
(Reproduced by permission of the
National Air and Space Museum
Smithsonian Institution)

November 15, 1952 1307 GMT Pyongyang, North Korea

This case is presented in Air Intelligence Information Report No. 52-350 dated 28 November 1952. The sighting took place from a T-6 aircraft over the central sector of the friendly main line of resistance. Following are excerpts from this report:

"At the time cited, Lt. (Layten S.) Bass was flying in the vicinity of CT795418 at 130 miles per hour indicated air speed, 10,500 feet, indicated altitude and in a northerly direction. The weather conditions were clear, with visibility from 15 to 20 miles. The sun at this time of day is in a south, southwesterly position from the incident area. Winds at 10,000 feet were generally from 240 degrees with velocities up to 20 miles per hour.

"Lt. Bass first noted the unidentified object moving north, apparently close to the terrain.... The object, initially dismissed by Lt. Bass as an aircraft, moved north approximately five (5) miles and then turned south and gained altitude....When the T-6 aircraft was at point #2 as indicated in the map overlay, Lt Bass again sighted the object and identified it as not being an aircraft. This sighting was to the right of the T-6, at an estimated 100-200 yards. Lt Bass described it as spherical in shape, silver in color and approximately 8-10 feet in diameter He further described the object as lacking any visable (sic) means of propulsion and having no protrusions of any sort. The surface was smooth. There was no audible sound.

When Lt. Bass saw the object, which was moving at approximately the same speed as the aircraft, he turned the aircraft towards it slightly at which time the object headed south and began climbing at a high rate of speed, turning into a left orbit around the aircraft (note 1). Lt. Bass called the object to the attention of Lt. Barnard who saw it as it went into it's left-hand orbit. (Lt. Barnard estimated the closest distance to the object as being from 500 to 1000 yards).

"Lt Barnard's description of the object was that it was a silver spherical object, approximately the size of a man's thumb nail held at extended arms length. (note 2) He further described the object as "shimmering" and having a "halo" which surrounded the object. The halo was only a hairline thickness. He further stated that the shimmering effect appeared to be more appareent when the object was between the aircraft and the general direction of the sun. The object appeared to Lt Barnard as being darker at the

> edges, as a translucent sphere would appear when held up to a light.

Other details were given. "The object was viewed against various colored backgrounds, the dark terrain, and blue sky and white clouds. It remained constant as to shape, size, and color." "Both sources state that the general motion of the object was steady and gave the impression of controlled flight. They stated that it did not appear to be moving on the motion of air current." "The pilot had flown 54 missions in this theater and the observer had flown 48 missions."

The preparing officer (1st Lt. James F. Power) considered both observers to be "compitent (sic) observers and sightings reported by them are considered reliable." There was no ground radar contact other than that of the T-6 aircraft. He also stated that the object "could have been (a) 350 gram (white latex) weather balloon" released from the Army's 1st Artillary Observation Battalion, a light plane unit located at Tokchong, CS3089, southwest of the location where this UFO was sighted. A balloon of this size was launched at 1257I on 15 November about 50 minutes earlier. However, Brigadier General Charles Y. Banfill, Deputy for Intelligence stated: "The 350 gram balloon released 50 minutes previous to the sighting would not be expected to be at any visible altitude at the time of the incident unless it were leaking, (which would preclude the rapid climb described). There previously have been a number of unidentified sightings in the CT and DT grid areas of Korea. Of these, at least 5 sightings have tentatively been identified as enemy balloons. Vectoring an aircraft on a free balloon is actually difficult, and false impressions of rapid movement can easily be made. From the available information it appears that the object was probably a balloon, and possibly one launched by frontline enemy troops." (See Appendix 1 for generally representative sizes and weights of U.S. balloons for comparison)

Signed: Charles Y. Banfill

Notes:

1. The reported behavior of the UFO is similar to that described above for October 16, 1952, less than one month before! This apparent reactivity by the UFO is intriguing.

2. This is approximately equivalent to an angle of 1.8 degrees arc.

Feb. 7, 1953 Korea

An F-94 pilot sighted a bright orange object both visually and on radar. It changed altitude and then accelerated away from the jet airplane at high speed. (Hall, pg. 21, 1964).

March 20, 1953 1200 Korean Time Pork Chop Hill, Korea

William Morgan, USAF sighted a UFO near Pork Chop Hill. No further information is available on this case.

April 2, 1953 1734 Korean Time NW Korea

Project Blue Book files contain a reference to a pilot sighting of a UFO in the Northwestern region of Korea. No other information is available on this case.

April 19, 1953 1300 Korean Time Seoul

"A white shiny round object was seen over Seoul at approximately 1 p.m. today (Sunday). Aerial observers in two separate planes flying routine reconnaisance mission observed a white, rounded, delta-shaped object. It was estimated to be five to seven feet in diameter. The observers had no idea of it's depth or thickness." It was travelling between 60 and 80 mph in a vibrating motion. The course of flight was north-northeast to south-southwest over enemy territory."

The object had no visible means of propulsion and sometimes flew only 100 feet above the ground. The *Aerial Phenomenon Research Organization* (APRO) Bulletin article citing this case stated, "The fliers, four Americans on reconnaisance flight, said it appeared to be round, only 10 feet in diameter and about 25 miles northeast of Panmunjom. (APRO Bulletin, vol. 1, issue 6, May 15, 1953). This sighting occurred north of Pork Chop Hill and Old Baldy Hill where heavy fighting had occurred in the previous several days. A ground observer said that "other luminous objects travelling at supersonic speeds of 800 miles per hour, had been observed in the Baldy-Pork Chop area and tracked on radar..." within several days of this sighting (note).

Three of the four aircrew were Lt. Julius Morgan (pilot); Lt. James O. Rymus (observer), and Lt. Jack E. Myers (observer).

Note. This is the same area as the event described in chapter 2.

December 26, 1953 1625 Korean Time CS 4859 North Korea

Project Blue Book files contain a reference to a five minute-long UFO sighting over North Korea. A two page teletype report provided the following information. Lt. Barr was piloting a T-6 aircraft from K-16 to K-47 at 1625 Korean time when he noticed an object crossing his course about three miles away. It was in a shallow dive (10 degrees) and travelling about as fast as a fighter airplane. Very soon its speed began to increase to an estimated 1,000 mph. Curious as to what this strange object could be he changed course to 45 deg, maintaining 3,000 feet altitude. He was travelling at 140 mph. Suddenly the object changed its course and turned toward his airplane. Lt. Barr "...started to circle in pursuit. The object was slightly lower and was turning inside of Lt. Barr... (who) approximated the radius of his turn at less than a thousand feet." The turn he was flying was not so tight as to cause a stall at 120 mph. "The object continued to turn inside of Lt. Barr's path so he widened his turn in order to make a straight pass." He lined up on the object at approximately 130 mph and started to close on it. The object then went straight up several hundred feet. The witness then decided to bank right in order to close or to see if the object would follow any given pattern. Lt. Barr last saw the object travelling toward the sun at about 140 mph.

The object was kite shaped, i.e., "a diamond shape with one short end"; it came to within 600 to 800 feet of his airplane during this encounter. It was estimated to be about six feet long and 18 inches thick. "The object looked to be higher on the outer edges and from the rear looked like a flat V. From the side, the object looked like a shallow triangle. It was orange and seemed translucent, at least on the outer edge. At one time the pilot saw "...what appeared to be sparks in the air a short distance behind the object."

Winter 1954 Pohang

John A. Potter, Marine Corps weather observer, sighted a formation of about seven discs moving in a side-to-side oscillation. (Hall, pg. 31, 1964).

Chapter 4

Ground Observers Report UFOs

This chapter presents a number of UFO sighting reports obtained from project Blue Book files which involved observers on the ground. That such sightings continued long after the Korean War is shown by an interesting case which took place in 1974 which was reported by Stringfield (1977).

April 5, 1952 Midnight Okinawa

This brief summary report was found in USAF intelligence report 131-52 dated 22 August 1952 (pg. 3). It is included here only to indicate that Korea was not the only place in the far east where these strange events were taking place. Okinawa was the location of a U.S. military base about 600 miles south of Korea. Many other sightings were also made from Japan. Following are the salient details.

A single small luminous UFO was observed flying at about 20,000 feet altitude approaching Kadena AFB, Okinawa where the eye witnesses, all flight crew members, were standing. At first it flew in a straight and level path approaching from the north but then it began to maneuver over the base in violent high speed zig-zagging flight. It stopped momentarily and then maneuved around. Then it shot up out of sight heading west. It was in sight for about ten minutes. The white steady light had no definite shape but was larger than a star. It did seem to become larger as it approached or smaller as it departed. It made no noise and had no trail, sparks, or flame. Its speed was estimated to be faster than any conventional jet aircraft, estimated at 1,000 knots.

April 30, 1952 1800 Korean Time Battlelines

"Two white, silvery objects, one behind the other, rumbled high in the sky over a valley in a frontline sector at 6 p.m., April 30th, heading south-east into South Korea," stated a Stars and Stripes article by Sgt. Bill McCorkle (April 1952 issue). He went on to point out that six members of the third platoon, Heavy Mortar Company, 180th Regiment, saw the objects. PFC H. B. Webb, a jeep driver, said that their speed was thought to be about 900 to 1000 mph. "A trail different than jet streams appeared behind them," he said.

PVT. Roland Jones, another eye witness said, "Jets passed over right before the objects, only they were Sabres going north. They were much slower than the southbound objects we saw." Mortar platoon leader, 2nd Lt. Bill Smith said, "They didn't have wings but they sounded similar to jets with their rumbling sound. But they weren't, I know. They were going way to (sic) fast.. and they started down, tumbling to earth, as they got out of view. Their trajectory before falling was flat and that proves they weren't mortars. They weren't jet gas tanks jettisoned off either as some theorized, as their trajectory knocks this idea too."

Note: Later, two holes were found in the ground in a reserve area. Each hole was about 25 feet apart, two feet wide, and seven and one-half feet long. Some felt that the "fresh holes" were caused by Chinese "Katyusha" rockets that had been used occasionally at the front. While it is possible that this or another type of enemy rocket that could fly horizontally was what was seen it isn't likely that this is what produced the two holes. No rocket casings were found and the size and shape of the holes were *not* those of a rocket explosion.

May 31, 1952 0400 Korean Time Korea

This ground radar site report was found in an Air Intelligence Information Report, dated 4 June 1952 prepared by Capt. John H. Cummings, 6004th AISS APO 970. An airman on duty at post #6 heard his field telephone ring. A guard at post #4 was calling to report seeing a "bright spot in the dark sky to the northeast... approaching fast." The object was oval shaped and about the size of a fighter aircraft. It was first seen at an estimated altitude of about 3,500 feet due north. It descended slowly to about 2,600 feet where it stopped for several seconds. The observer stated that it looked like a disc at this point. It had a sound like a pulsating jet engine at idle. Then it started moving east for about a mile with jerky movements and then stopped again momentarily (note). It reversed its direction abruptly and headed west for a short distance to almost its original position (when first seen). It stopped and hovered for a moment, then reversed direction, moving toward the east in a shallow climb and then a steep climb (about 45 degree angle) to the north and disappeared in from 3 to 4 seconds time. (also cf. Gross, pg. 90, 1982)

Drew Pearson, the noted newspaper columnist wrote: (The Washington Post; March 7, 1952) "The Pentagon has hushed it up, but radar units have actually tracked 25 "flying saucers." Each was sighted by at least one eyewitness and also picked up on the radar screen."

Note.

The description of the UFO given in Chapter 2 also included very erratic movements and a deep throbbing sound when the soldier's rifle bullet struck the aerial object.

June 25, 1952 1123 GMT Sawon

A wire to FEAF from CG FEAF, Tokyo, Japan dated 25 June 1952 referred to a ground sighting of UFOs at location K-14 (note 1) at 1123I. It looked like a "coin", dull silver and appeared to be spinning in a CW direction. Its diameter was estimated to be seven (7) times its thickness. Its edges were dark blue. "It was reported for shooting a course at an estimated speed of 300 knots, alternately revolving 360 degrees in seven (7) seconds and then proceeding on a spinning course for approximately 12 seconds. The object approached K-14 from the north, paused momentarily, and then rose straight up for 10 seconds, the exact altitude unknown, and then disappeared into the sky. A few seconds later this object was again observed by a second person at the same point of observation. The unidentified object which was seen proceeding at a rapid rate of speed from the south to the north at an approximately 45 degree angle, pausing and then returning in the same manner, each time passing in front of the sun and travelling an equal distance to each side of it. The time required to progress in one direction was three seconds." (Ibid.) (note 2)

Notes:

1. Air base K-14 was located near Sawon, just south of Seoul.

2. This case contains some similar elements to the May 31, 1952 incident.

July 10, 1952 **North Korea**

Sailors on a Canadian destroyer sighted two shiny revolving discs race across the sky. Ship radar indicated their distance was seven miles. (Hall, pg. 160, 1964)

September 18, 1952 2230 Korean Time **Western Korea**

Gross (1986) notes a report in which "...six spark-throwing cartwheels were seen over the front lines by U.N. ground troops" for about 15 minutes at about 10:30 p.m. local time. Each object seemed to be about 18 inches in diameter and moved in a fifteen foot circle. (Lorenzen and Lorenzen, pg. 48, 1969)

October 19, 1952 **Front battle line**

Gross (Ibid., pg. 77) recounts another similar sighting as above where "cartwheel-like objects throwing sparks flew in a 15 foot orbit above the front on the west side of the military line of resistance (MLR)." J.A. Lajoie of "A" Corp. dismissed the flare possibility and suggested that the object resembled Fourth of July fireworks (Anon, 1952).

January 24, 1953 0650Z **Suwon, Korea**

This report appeared in USAF Intelligence Report IR-1-53 dated 26 January 1953. The observer, a Staff Sargeant, was on the ground looking upward at about 45 degrees elevation toward the east when he sighted two round objects for between three and five seconds. They appeared to glow a whitish silver color and were convex with a dark line or shadow traversing the base of each one. His estimate of their size was about twelve to fourteen inches in diameter and at a distance of about four miles away from him. As he continued to watch them they flew in trail in a gentle climb. They seemed to be travelling "considerably faster" than an F-86. No propulsion features, trails, exhausts or sounds were noted. They disappeared either into or behind a nearby cloud.

The following day the witness was interrogated again and changed his description of the objects. On the basis of this evidence the report states, "His judgment of speed, distance and elapsed time of observation are estimated to be unreliable." The report does not include any facts about what he said on the second day, however.

The official explanation for this sighting was PROBABLY AIRCRAFT because two F-86 jets had flown over the field at lunch time. "During these maneuvers, the aircraft at times appeared only as silver dots in the sky. If an observer did not follow these aircraft through the complete maneuver, it would have been almost impossible to tell what they were." (Ibid., pg. 3 of 3)

February 10, 1953 0025 Korean Time Cho-do, North Korea

Air Intelligence Information Report 62-3-53 dated 6 March 1953 contained the following ground sighting details. A Marine Sargeant, John Muciek, on communications watch at the communications tent of the USMC garrison at Cho-do, saw an object which looked like a falling star. The light was a steady white and moved in an approximately level path. Then "it stopped abruptly, and moved back and forth in that area (KC-5555)." The light was replaced by a red light and "...from then on, at intervals of approximately three (3) seconds, the light flashed alternately red to white." Sargeant Muciek then called to Corporal Hubert Hicks inside the tent to come out, which he did. He also saw the light. After another minute or so the object began to move away to the southwest. At no time was sound heard from the object. Then Sargeant Muciek phoned the sighting into the USMC Command Post.

The preparing officer was Donald Nichols, Detachment Commander. He noted that the next day he spoke with a Major Pak, KMC Commander, who said that he too had seen "...a flying saucer at approximately the same time with approximately the same phenomena as described by the two (2) Marines." (Ibid., pg. 2 of 3). The official USAF explanation on the summary card for this highly maneuverable light was ASTRO (VENUS). Nonetheless, Col. George D. Hastings, Deputy for Intelligence concluded that the witnesses saw a lighted balloon since "...a number of reports have been received of enemy activities with balloons...". (Ibid., pg. 3 of 3)

April 29, 1953 1700 - 1705 Korean Time Eastern Pacific Ocean
28 deg 14 min N 166 deg 5min W

A small metallic appearing object the approximate size of an F-80 fighter aircraft was sighted from the USNS General A. W. Brewster enroute to FEAF from San Francisco. It was about 5:00 pm local time when officers on the deck sighted the UFO at the 3:00 o'clock relative position and 2,000 feet in the air, slightly below the clouds. It was about 1/2 mile away from the ship. The object seemed to "remain stationary for a few seconds, then turn at right angles towards the ship, gain altitude and travel in an arc to a position of 5 o'clock disappearing through the clouds aft of the ship."

"The object did not perform with the smooth control movements of any known aircraft. Rather its controlled actions were erratic with some turns at 90 degree angles and near vertical climbs. The movement from 3 to 5 o'clock was in an arc." The witness who reported this sighting estimated the speed of the object to be faster than a jet fighter aircraft.

The End

References

Anon., Associated Press, Tokyo, Japan, October 19, 1952.

Anon., *Aerospace Science: History of Air Power.* Air University Press, Air Training Command, Maxwell A.F.B., Alabama, 1986.

Braybrook, R., *Supersonic Fighter Development.* Haynes Publ. Group, Somerset, England, 1987.

Gross, L.E., *UFO's: A History January-May 1952.* Privately published, 1982.

Gross, L.E., *UFO's: A History 1952: September - October.* Privately published, 1986.

Haines, R.F., Commercial jet crew sights unidentified object-Part I. *Flying Saucer Review*, Pp. 3-6, March 1982; Part II. Pp. 2-8, April 1983.

Haines, R.F., and J.F. Vallee, Photo Analysis of an Aerial Disc Over Costa Rica. *Journal of Scientific Exploration*, Vol. 3, No. 2, Pp. 113-131, 1989.

Haines, R.F., and J.F. Vallee, Photo Analysis of an Aerial Disc Over Costa Rica: New Evidence. *Journal of Scientific Exploration.* Vol. 4, No. 1, Pp. 71-74, 1990.

Hall, R.H., *The UFO Evidence.* National Investigations Committee on Aerial Phenomena (NICAP), Washington, D.C., May 1964.

Jackson, R., *World Military Aircraft Since 1945.* Charles Scribner's Sons, New York, 1979.

Jacobs, D.M., *The UFO Controversy in America.* Indiana Univ. Press, Bloomington, Indiana, 1975.

Lorenzen, C. and J. Lorenzen, *UFOs The Whole Story.* New American Library, Times Mirror, 1969.

Mahon, J.K., and R. Danysh, *Infantry Part I: Regular Army*. Army Lineage Series, Center of Military History, United States Army, pp. 501-503.

Marshall, S.L.A., Bayonet Charge. *Combat Forces Journal I*, Pp. 22-32, 1951.

Michaelis, M., and B. Davidson, This we learned in Korea. *Collier's*, Pp. 13-15, 38-39, 42-44, August 1951.

Momyer, W.W., *Air Power in Three Wars*. (Lt. Col. A.J.C. Lavalle and Maj. J.C. Gaston (eds.), Pp. 358, Privately published, 1978.

Morse, J.L., (Ed.), *Funk and Wagnalls Standard Reference Encyclopedia*. Standard Reference Library, New York, vol. 15, 1969.

Nowarra, H.J., and G.R. Duval, *Russian Civil and Military Aircraft, 1884-1969*. Fountain Press, London, 1971.

Pearl, J., Wolfhounds: Saga of the 27th Infantry Regiment. *Saga*, Pp. 48-50, 69, 71-72, August 1963.

Pullen, R., et al., (Eds.), *The Tropic Lightning in Korea*. Love Publ., Atlanta, GA, 1954.

Stockwell, R.E., *Soviet Air Power-Supplement*. Pageant Press, Inc., New York, 1956.

Stringfield, L.H., *Situation Red, The UFO Siege!* Doubleday, New York, 1977.

UFO Information Research Center, Report 6601, Pp. 3-4 to 3-5, 1966.

Whiting, K.R., *Soviet Air Power*. Westview Press, Boulder, Colorado, 1986.

Appendices

1. Weather Balloon Characteristics

The following meteorological balloons were used by the Americans in the Korean War. This data was included in report IR-4-52, page 2 dated 7 June 1952.

Weight	Size	Color	Rate of climb*	Altitude
10 gram	18 inches	red or black	400	25,000 ft.
30 gram	24 inches	red, white, or black	600	60,000 ft.
100 gram	40 inches	red, white, or black	990	100,000 ft.
350 gram	12 ft	cream	1100	120,000 ft.
500 gram	25 ft	cream	1200	140,000 ft.

* ft. per minute

Times of Release

GMT	Local Time	
0600	1200 (noon)	
0900	1800	350 gram balloons flown
1500	2400	during daylight hrs. 750
2100	0600	gram balloons flown during darkness.

Thirty and 100 gram balloons are released from all USAF pibal stations. Ten or 30 gram balloons may be released from any station listed, whenever the ceiling is below 5,000 ft. Thirty and 100 gram balloons are occasionally released on an unscheduled basis. Five candle-power battery powered lights are sometimes attached to ten, thirty, and one hundred gram balloons during hours of darkness. "No instance of a balloon burning in the air has ever been reported in this theater." A chart of South Korea gives the 14 launching stations.

2. *Shapes and Sizes of UFO Reported*

Japanese lantern
Round
Globe (larger than Navy jet)
Oval (larger than a MiG)
Round black (circular) cloud
Coin
Large dark object (long streamers)
Perfectly circular (25 ft diam.)
Discs (7 total in group)
Cartwheel (giving off sparks)
Coolie hat with oblong ports
Ball
Cylindrical with a tail
50 foot diameter
Round
Disc (7:1 ratio and 15-20 ft)
Cigar (6-7 feet long)
Spherical (8-10 ft diameter)
Oval object (size of figher)

3. *Eye Witnesses Listed by Sighting Date*

Event Date	Name(s)
Spring 1951	Frances P. Wall
Feb. 24, 1952	R. W. Smith
Mar. 29, 1952	Brigham
April 30, 1952	H. B. Webb, Roland Jones, Bill Smith
May 15, 1952	McCarthy
May 26, 1952	Martin _ _ _ _ _ _ _
May 30, 1952	Robert J. Arblaster; William Leet; Andrew J.Roe
May 31, 1952	Hufberry
June 20, 1952	Bobbie Foster, Richard Francisco, Teddy L. Pittman, Ronnie A. McDonald
Aug. 9, 1952	Leo N. Nagrodsky
Sept. 16, 1952	C.B. Scott Jones
Sept. 18, 1952	J. A. Lajoie
Oct. 16, 1952	Hooper; O'Neill
Nov. 15, 1952	Layten S. Bass; Barnard
March 20, 1953	William Morgan
Winter 1954	John A. Potter

Note to Reader: If you served in the Korean War and know anything about UFO sightings you are urged to contact the author. Use the Publishers address: P. O. Box 880 Los Altos, Calif. 94023-0880.

4. Summary of Unusual or Provocative UFO Flight Maneuvers

Of major interest here are the flight dynamics of the UFO which appear to be *directly* related to the presence and/or behavior of the eye witness' airplane or the witness himself. For an aerial vehicle to be able to control its position relative to another vehicle rapidly it must possess (1) a large acceleration capability (which requires a large amount of energy, (2) a propulsion capability to direct the energy in a given direction efficiently, (3) a means of sensing the change in position of the other vehicle, and (4) a highly precise guidance and local "navigation" system. The airplanes and other aerial weapons which were used during the Korean War were mostly of World War 2 vintage and had only rudimentary capabilities in these four areas.

The dynamics presented here are abbreviated since they are presented in detail above.

Spring 1951 The object changed its position rapidly, immediately after the soldier's rifle bullet struck it.

Sept. 1950 UFOs maneuvered around the three fighter-bombers whose radars had become inoperative at that time. Objects then departed at a high rate of speed.

Fall 1951 Radar targets "had been circling the fleet; it (UFO) left the fleet and joined up on us a position behind our wingman."

Jan. 29, 1952 UFO flew parallel to airplane at first and then approached it at the same level. It stayed at a constant distance for about a minute and then departed.

Feb. 24, 1952 The object first approached the airplane from below. Then it levelled off at same altitude as airplane's and then turned left to intercept the observer's airplane after which it descended out of sight. (It did not look like an airplane).

May 15, 1952 UFO began a steep climb but at the top it then moved forward in level flight, wavering momentarily. It then descended and disappeared into the haze at an estimated 1000 mph.

May 26, 1952 An object followed the observers' jet interceptor. Subsequently, the UFO was in front of the jet with a positive radar lock-on. At about 6000 yards between the airplane and UFO (and closing) the object started to pull away to 26,000 yards in about 12 to 15 seconds (estimated at about 20-g acceleration).

May 31, 1952 A bright oval disc was seen from a ground observation post approaching fast. Then it descended slowly about 900 feet when it stopped several seconds; then it moved east with jerky movements, stopped again, reversed its direction 180 degrees flying to its original position. It hovered in one spot for a moment, reversed direction again in a shallow climb and then a steeper climb, disappearing from sight in 3 to 4 seconds.

May 31, 1952 As jet fighter turned and descended to intercept a UFO 6000 feet lower, it (UFO) began a left turn climb which silhouetted the F-94 against the dawn sky. After attempting two unsucessful passes on the object, the UFO began to accelerate away out of sight.

June 6, 1952 UFO travelled at about 300 mph in a straight line (while revolving) and then stopped above airbase K-14 momentarily and then rose straight up for 10 seconds after which it disappeared from sight.

June 20, 1952 The oval shaped UFO approached from the south, made a complete left-hand orbit of the target area and flew away to the east. The UFO outdistanced a pursuing aircraft.

June 23, 1952 The UFO descended "rather irregularly", flew at a fixed altitude for 6 - 7 seconds, approached the pilot's airplane and then disappeared into a cloud.

Oct. 16, 1952 An oncoming UFO on the right side and at the same altitude as the jet aircraft suddenly seemed to stop when abreast of the aircraft. Then, as the aircraft banked right toward it, it accelerated away at an estimated 800 mph.

Nov. 15, 1952 When first seen, the UFO was below the pilot and moving north about five miles. Then it turned 180 deg and headed south and climbed. When he banked toward it the perfectly spherical, silver UFO banked into a left hand orbit around the airplane.

Feb. 27, 1953 When the UFO was sighted (visually and on radar) it changed altitude and accelerated away from the jet airplane.

5. *Electromagnetic Effect Cases*

Sept. 1950 When a Navy airplane was approached by two large discs its radar was jammed and the radio transmitter was blocked by a buzzing noise each time a new frequency was tried.

Fall 1951 Over 14 (ground, airborne) radar sets tracked a UFO circling over the U.S. Naval fleet at speeds of from "slow" to over 1,000 mph.

May 26, 1952 Ground radar contacted a UFO and vectored a jet interceptor towards it. The radar operator onboard visually sighted the object which was subsequently locked on his airborne set.

May 31, 1952 The F-94's radar obtained a positive lock on the object while ground radar did not. It was first sighted from the ground.

July 10, 1952 Canadian destroyer radar contacted two UFO.

Aug. 9, 1952 Marine ground control radar contacted a UFO travelling at a high rate of speed.